WHEN CHAMPAGNE BECAME FRENCH

THE JOHNS HOPKINS UNIVERSITY
STUDIES IN HISTORICAL AND POLITICAL SCIENCE
121ST SERIES (2003)

Kolleen M. Guy
*When Champagne Became French:
Wine and the Making of a National Identity*

Christopher Forth
The Dreyfus Affair and the Crisis of French Manhood

WHEN *Champagne* BECAME FRENCH

WINE AND THE MAKING OF A NATIONAL IDENTITY

KOLLEEN M. GUY

THE JOHNS HOPKINS
UNIVERSITY PRESS
BALTIMORE AND LONDON

In memory of William B. Cohen,
1941–2002—Mentor, Scholar, Friend

© 2003 The Johns Hopkins University Press
All rights reserved. Published 2003
Printed in the United States of America on acid-free paper
9 8 7 6 5 4 3 2 1

The Johns Hopkins University Press
2715 North Charles Street
Baltimore, Maryland 21218-4363
www.press.jhu.edu

LIBRARY OF CONGRESS
CATALOGING-IN-PUBLICATION DATA
Guy, Kolleen M. 1964–
When champagne became French : wine and the
making of a national identity / Kolleen M. Guy.
p. cm
Includes bibliographical references and index.
ISBN 0-8018-7164-6 (hardcover : alk. paper)
1. Champagne (Wine)—History. I. Title.
TP555 .G89 2003
641.2'224—dc21   2002008103

A catalog record for this book is available
from the British Library.

Contents

ACKNOWLEDGMENTS	ix

One
Introduction ... 1

Two
Consuming the Nation: Champagne Marketing and Bourgeois Rituals, 1789–1914 ... 10

Three
Industry Meets *Terroir*: Champagne Producers in the Marne, 1789–1890 ... 40

Four
Resistance and Identity: Cultivation Methods and the Wine Community, 1890–1900 ... 86

Five
Boundaries: The Limits of the "True" Champagne, 1900–1910 ... 118

Six
Revolution and Stalemate: The Revolt of 1911 ... 158

Seven
Conclusion: Champagne and Modern France ... 186

APPENDIX	197
NOTES	199
BIBLIOGRAPHIC ESSAY	231
INDEX	239

A gallery of photographs follows page 50.

Acknowledgments

This book began as a doctoral dissertation at Indiana University, where I benefited from the sound advice and good counsel of William Cohen, George Alter, Michael Berkvam, and James Diehl. George Alter helped me to concentrate on some key issues that became central to the book. Through the years, William Cohen's incisive criticism, constant support, and enthusiasm have continued to inspire me. He has taught me the craft of the historian and been the model of a conscientious teacher and outstanding scholar. While in Bloomington, I also had the good fortune to serve as an editorial assistant at the *American Historical Review*. The office on Atwater Street was more than a place to work, it was a place of extraordinary intellectual exchange and great friendship. I thank my former colleagues at the journal, who taught me much about good writing and the value of intellectual exchange across the various fields within the discipline of history.

I owe a debt to librarians and archivists in France and the United States. Most of the research for this book was done in France at the Archives nationales, the Bibliothèque nationale, the Bibliothèques municipales in Reims and Châlons-en-Champagne (formerly Châlons-sur-Marne), and the Archives départementales de la Marne in Châlons-en-Champagne. I would like to thank the directors and staffs of these institutions for their support with this project. The staff of the Archives départementales de la Marne, where I spent the largest portion of my time, provided not only professional assistance but also a collegial atmosphere in which to work. A special note of thanks to Louis Bergès, director of the Archives départementales de la Marne, for permission to use some of the illustrations that appear in this book. The public archives of France are rich, but without access to private archives, my research on the champagne industry would have been hindered greatly. A special word of thanks to Yves Lombard, former general director of the Syndicat des grandes marques de Champagne, and his staff who not only provided unlimited access to some of the most important papers for my research but also made my stay in Reims enjoyable by providing a cozy place to work and a tasting of some of the best wines in France. On this side of the Atlantic, the library staff at the University of Texas at San Antonio have been instrumental in providing me with interlibrary loan materials.

I feel a special gratitude to a number of individuals who have provided intellectual and material assistance to me during the course of this project. Thanks to my friends Catherine Poupier and François Girault for their ongoing support, and to my many other friends in Orlèans for their incredible efforts to make me feel that the city is home. My gratitude also to the writer François Bonal for sharing his considerable knowledge of Champagne. The historians Christophe Speroni and Martin Bruegel took an active interest in this project and helped me to make invaluable contacts. My colleague and friend Steven Zdatny has offered moral and intellectual support from the very beginning of this project. I must also express my deep appreciation to Barbara Sciacchitano, William Beik, and David Blight, who guided me from very early on in my academic career and continue to influence me in ways both large and small. Closer to home, I have benefited from the indispensable advice and steadfast support of my many students, colleagues, and friends at the University of Texas at San Antonio. I would like to give special thanks to my colleagues Antonio Calabria, Brian Davies, David Johnson, Patrick Kelly, Yolanda Leyva, James McDonald, Juan Mora-Torres, and Wing Chung Ng, who read various parts of this manuscript and offered helpful comments and useful suggestions. Since I have presented parts of this work at conferences and have explored some of my ideas in other publications, I wish to thank all of those who generously provided comments and suggestions along the way, particularly Warren Belasco, Thomas Brennan, Ann Carlos, Herrick Chapman, W. Scott Haine, Steven L. Harp, Gary Kates, Leo A. Loubère, Leslie Page Moch, Roderick Phillips, Donna Ryan, and Philip Whalen.

I gratefully acknowledge the financial support that made this research possible. During the initial phase of dissertation research and writing, I was generously supported by the Fulbright Foundation, a Gilbert Chinard Fellowship, an Indiana University College of Arts and Sciences Dissertation Year Fellowship, and a Center for International Business Education and Research (CIBER) Award from Indiana University. A Faculty Research Award, a Department of History Summer Faculty Grant, and a College of Liberal and Fine Arts Summer Fellowship from UTSA provided me with valuable time in France to undertake additional research and to finish the manuscript. The Herman Krooss Dissertation Prize from the Business History Conference provided financial assistance during the early stages of writing. At a time when funding for education and research is increasingly scarce, I was particularly lucky to find such generous support. I would also like to thank Bob Brugger, Melody Herr, and the editorial staff at the Johns Hopkins University Press for their tremendous patience and unflagging enthusiasm.

Deep thanks are due to my family and friends for their moral and material support. Through thick and thin, my mother, Kathleen Lange and my brother, Jim Guy, have provided me with heavy doses of emotional support. James L. Guy, Sr., my father, and Bettye Bishel, my mother-in-law, did not live to see the completion of this project, but I know that each, in their own way, would have been very proud. I imagine no one would have been prouder to see the publication of this book, however, than my dearest friend, Jim Peters, whose untimely death took away my best critic. I would also like to acknowledge the love and support of Bob and Amy Bishel and Brian and Paula Davies, as well as of my Bloomington family, Masha, Clayton, Anna, and Nina Black. Brian Davies and Clayton Black, both Russian historians, have taught me the real meaning of the term "comrade." My friendship with Cathy Haubtmann has deepened over the many summers that we have traveled across France discovering the *terroir* together. She has provided me with a home in France from which I can research and write. Her generosity and loyalty are an incredible gift.

The irony of finishing this book is discovering that the acknowledgments are the hardest to write. Somehow I believed that thanking those who helped during the lengthy process of researching and writing would be a relatively simple task. Now, however, I find my words of thanks woefully inadequate. This is particularly true as I search for the words to thank my son Eric Guy Bishel and my husband, Bill Bishel. Eric has given up his Mama to France every summer since he was born. He has tried very hard to be supportive and has been a constant source of good cheer. As we await his new sibling, he continues to try to think of ways to get Mama back to France. I am blessed with a remarkable little boy. As for my husband, Bill, we were friends and colleagues long before we became husband and wife. Over the years, Bill has shared his love of history while helping me to develop my critical judgment. I owe him a great intellectual debt. Although I owe him so much, I know that he would never try to collect. And, for this, I owe him my biggest thanks. It is to Bill that I dedicate this book.

WHEN CHAMPAGNE BECAME FRENCH

INTRODUCTION

> La belle France! ... What wine! What diversity, from bordeaux ... to sparkling champagne! What variety of white and red, from Petit Mâcon to ... Aÿ mousseux!
>
> FRIEDRICH ENGELS,
> "Seine und Loire" (1848)

Champagne. The word has found its way into languages far removed from French. People who have never seen, let alone tasted, French sparkling wine use the word as an image. Writers, painters, and musicians, from eighteenth-century *philosophes* to twentieth-century jazz singers, contribute to the ongoing invention of the image by using the wine to denote social status and, more significantly, the glories of France.

Within France, champagne has been seen as an embodiment of the national spirit. The pioneers of French gastronomy in the early nineteenth century, such as Anthelme Brillat-Savarin and Alexandre Grimod de La Reynière, associated the production and consumption of food and wine with the fate of the nation.[1] The medicalization of society over the course of the century gave the early "science" of gastronomy new authority. Scientists and popular writers linked champagne with a unique French personality. "The French, merry and blithe, are much like their wine of Champagne," the famous doctor Léandre-Moïse Lombard concluded in his monograph *Le Cuisinier et le médecin* (The Cook and the Doctor).[2] "Champagne is the French wine par excellence," Adolphe Brisson declared. "The wine resembles us, it is made in our image: it sparkles like our intellect; it is lively like our language."[3] The French people and their sparkling wine were seen as sharing an animating element.

Popular magazines and books on wine in the twentieth century continued to echo this opinion, calling the discovery of champagne by Dom Pierre Pérignon, a monk at the abbey of Hautvillers between 1668 and 1715, one of France's greatest achievements.[4] Since 1900, moreover, this sense of champagne's importance to the nation, to French collective identity, has been reinforced by the actions of the French government, which aggressively protects the appellation as a part of the national patrimony. "Champagne" is now preserved as a trademark of France within the European Union. Champagne is "our patrimony and our collective trademark," a spokesman for the French champagne maker's association, the Syndicat des grandes marques de Champagne has stated.[5] Few in France appear to disagree.

Champagne, some would argue, is "rooted" in soil and history, connected with place, transcending time, and offering a genuine experience of France. Consumption of champagne provides natural access to an authentic, organic France through the intermediary of French *terroir*. A term with no precise equivalent in English, *terroir* has generally been used to describe the holistic combination in a vineyard environment of soil, climate, topography, and "the soul" of the wine producer.[6] Terroir was (and often continues to be) seen as the source of the distinctive wine-style characteristics at the heart of fine champagne. Much like the nation, champagne and its terroir are believed to possess eternal, natural qualities. The wine can be seen as an objective manifestation of the French "soul," the guardian of supreme spiritual values. "Champagne remains a symbol, profoundly rooted in French culture," notes one contemporary author, "it magnifies the virtues produced from good peasant soil and the *esprit voltairien*."[7]

Historical narratives of wine and the champagne industry of France reinforce this sense of timeless authentic Frenchness by chronicling the classical origins of French wine and highlighting its various golden ages. In his classic *Histoire de la vigne et du vin en France des origines au XIXe siècle*, Roger Dion writes that "our elite vineyards," defined as those areas made up of "fine, quality vines" where winemakers use methods elaborated over centuries of practice, are one of the "most glorious expressions of our civilization, bequeathed to us from ancient Greece and Rome."[8] For Dion as well as many other French historians, to write the history of wine is to write the history of the French people, a history grounded in an ancient past and a timeless terroir, which serves as the repository for the accumulated historical memory of France.[9] Wine produced from that terroir appears as part of the "rich legacy of remembrances," to use the famous words of Ernest Renan, that forges the solidarity of the French nation.[10]

Champagne is the subject of the ultimate chapter in Dion's magisterial survey of the history of the vine and wine in France. Its creation in the eighteenth century marked the advent of modern prestige viticulture, and in the nineteenth century its production developed into a large-scale commercial undertaking, "new to the vinicultural history of France."[11] For much of the period he studies, Dion notes, what is commonly thought of as champagne—sparkling white wine bearing the regional appellation—did not exist. Indeed, the regional still wines of Champagne were not associated with the name "champagne" until the eighteenth century and, even then, the appellation was not widely known.

The natural environment of *la côte de Champagne*, the terroir of the region, as Dion foreshadowed in his introduction, made the region "predestined" for fine wine production.[12] The modern *vin mousseux* of Champagne that emerged in the eighteenth century was part of a longer narrative rooted firmly in antiquity. Champagne originated in an ancient "cult of the vine" driven in France from the sixteenth through the nineteenth century by an *amore patrio*. Local wine growers nurtured an "ancestral tradition" of quality still wine cultivation out of a "patriotic obligation."[13] This fidelity to ancestral traditions of quality wine production built the foundation for the innovative winemaking experiments of Dom Pérignon and the other key "creators" of champagne.[14] The story of the emergence of champagne and its unparalleled reputation in the nineteenth century is, in Dion's narrative, the logical culmination of French history.

In this context, why would anyone bother to reconsider *when* Champagne became French? Indeed, champagne, as a good associated so intimately with the national history of France, has an authority and legitimacy not afforded most commodities.[15] French wines, as demonstrated in Dion's work, are approached much like the French nation itself. Eternal virtues or qualities that are attributed to both France and its wines can, however, disguise what are social and cultural constructs in natural attire. Emphasis on objective factors such as soil and climate, which are at the heart of terroir—used to distinguish the excellence of *crus* and the material profile of countries—often blurs the importance of systems of social values in the invention of both wine and nation. Analogies between wine and nation dominate, creating an artificial determinism within French winegrowing history. A certain circular logic follows from this history, making it possible to use the same words to describe both the qualities of Frenchness and the qualities of French wines. Wines and national identity become so intertwined that it is difficult to invoke the one without eliciting the other.

When Champagne Became French addresses both wine and nation. By examining the historical relationship between champagne, social distinction, and French national identity, this book seeks to go beyond the teleological and self-referential logic that is at the heart of much of the historiography of the French wine industry. Both champagne and French winemaking entered a new era in the nineteenth century with the growth of markets for prestige or luxury wines and the advent of large-scale wine production. Thomas Brennan's masterful study *Burgundy to Champagne: The Wine Trade in Early Modern France* confirms this development, chronicling the half century—roughly from 1775 to 1825—when production of bottled wine became more sophisticated and the export trade became big business.[16] By the mid-nineteenth century, sales of champagne were calculated in the millions of francs, and it quickly became one of France's most profitable exports.[17]

Annual champagne sales climbed rapidly after 1870, topping 21 million bottles sold abroad and 4 million bottles sold at home by 1890. New techniques of producing, selling, and distributing champagne emerged in tandem with the increasing dependence of the nation-state on the opinions of ever-broader elements of the population. Luxury prestige wines became a part of the new, mass consumer culture and national consciousness. It was in this era that the French began to discuss champagne's place within their national culture. Defining champagne as French became a highly contentious issue both at home and abroad.

The late nineteenth century was a time of rapid change, including the beginnings of the modern revolution in consumption. Social groups and their environments were dramatically transformed, leading them to search for new devices to ensure or express social cohesion and identity and to restructure social relations. Champagne was central to this process. Used to delineate social boundaries, champagne consumption became a basic ritual for membership within social groups. As an integral part of numerous traditions and rituals, champagne became a subject of mass culture, a centerpiece of bourgeois society. Whereas some of these traditions were consciously invented and constructed, others evolved more informally with varying degrees of deliberate construction. Whether invented or evolved, these traditions, rituals, and images became a part of what the French sociologist and cultural critic Pierre Bourdieu has termed "cultural capital."

Champagne's investiture with cultural capital resulted from both its linkage to and reinforcement of France's burgeoning reputation as the preeminent capital of the bourgeois world and the wine's own evolving importance as a transnational marker of social distinction. The symbolic significance of the commodity and its use in the ritual symbolism of diverse societies invested those who controlled its production and distribution with a vital element of economic and social control. No single group, agency, or institution had the power to create or consolidate the rituals and images surrounding champagne, but the peasant vine growers (*vignerons*) and merchant-manufacturers (*négociants*) who produced champagne in the department of the Marne found themselves uniquely positioned to profit from cultural attitudes about the wine and its consumption.

With the rapidly evolving consumer culture of the end of the century, "old" images of champagne went through frequent modifications, eliciting efforts by these regional producers to repress unwanted traditions or symbolic readings and encourage "correct" ones.[18] These correct images were also in the interest of the new French republic, which sought to promote France's economic position as a major supplier, not only of quality wines and agricultural products, but also of luxury manufactured goods. By the eve of World War I, some were declaring champagne to be no less than a "Citoyen du Monde entier."[19] The universal citizen, however, was associated with the glories of the French nation. Champagne, for all its worldliness, emerged as a symbol of France.

As part of a new range of social symbols for the aspiring elite, champagne could readily be exploited by those who supplied it. Yet this symbol of the French nation was a regional wine, produced from grapes grown by local vine growers in areas of what had once been the province of Champagne—carved up by the Revolution into the four separate *départements* of the Marne, Haut-Marne, Aube, and Ardennes—and manufactured and marketed by private business interests. Being a regional product but promoted as a national good in advertising and marketing spectacles, champagne gave the community that controlled its production a singular importance within the nation. Beginning with debates over local treatment programs for vine diseases in the 1880s, regional concerns

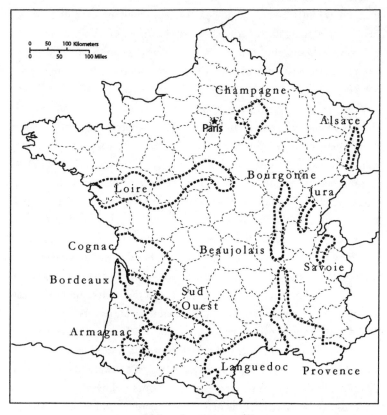

MAP 1. Major wine regions of France

of the peasant grape growers and merchant-manufacturers were increasingly presented as national concerns. By 1900, the champagne-producing community of the department of the Marne had developed a rhetoric of national identity that promoted its own interests as those of the nation. Its ability to successfully mask local interests as national concerns convinced government officials of the need to protect champagne as a national patrimony at both the national and international levels.

Protecting the national patrimony, however, was not a simple matter. Champagne, like the French nation it represented, was perceived as having eternal qualities that disguised social and cultural constructions. Negotiations for protection at both the national and international levels brought into the open the contested nature of cultural representations of champagne and, by extension, the French nation. Efforts by groups within the regional industry and imitators, both in France and abroad, to profit from the success of champagne on the world market raised a number of issues about the use of regional appellations, the counterfeiting or imitating of brand names, and the delimitation of French wine regions. Challenges from abroad merged with more generalized concerns about French economic performance and French power and prestige in the decades leading up to World War I.

Attempts by the French government and the champagne producers to resolve particular issues affecting the regional wine industry took on a national urgency. As a marker of French identity, both the wine and the region were believed to be under siege, not only from abroad, but also from within, as a result of crises between 1900 and 1911 involving fraudulent production and falling prices. The national response was to create new forms of protectionism designed to legitimate and limit access to the national patrimony. Protective legislation at the local level brought to the surface the conflict that had emerged within the champagne industry, where the vignerons and négociants were engaged in complex social and economic relationships. Each side promoted its positions, its concerns, as those of the nation, demanding a national response. State efforts to appease the various "protectors" of the patrimony stalled in 1911. Arguing that national honor was at stake, the peasants of the champagne industry took to the streets in a bloody revolt, which ended only after nine months of armed confrontation and military occupation. A little over a century after the French Revolution, the Champagne region was the site of fraternal discord over some of the most fundamental assumptions about the French nation.

Events in the wine industry of Champagne demonstrate how private forces and the rural periphery interacted with urban public institutions at both the national and the international levels to shape French identity. Historians of modern France have depicted state efforts after 1870 to integrate the separate *pays* that made up the rural world into the nation as a process akin to colonization. The history of champagne suggests, however, that in questions of nation-building and forging a national consensus, local forces and private companies were pivotal, sometimes working in conjunction with the state but, as the periods of contention highlight, sometimes acting in open opposition to it. Private companies and local peasant organizations promoted their regional specialty, champagne, as a national good, integral to the common imagined past actively promoted by the state. Rather than a construction of "France" as a nation that was imposed by Paris on the periphery, the regional wine community helped

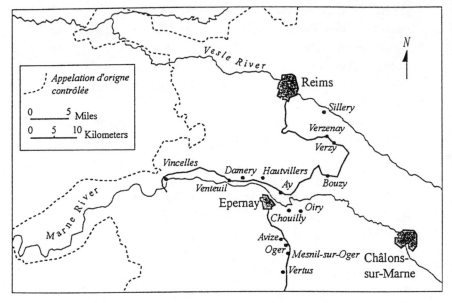

MAP 2. Major towns and cities of the Champagne region

to manufacture a common culture in which local traditions became national and local regions became national territory.

Champagne is asserted to have been granted the first legally recognized *appellation d'origine* in 1908. Today, champagne differs from other sparkling wines by an obligation to conform to certain industry production standards, now part of French law. Current laws regulating the production of champagne are as complex as they are strict. Rigorous controls are used to assure that only grapes from the geographic area delimited as "la Champagne" are used in blending and to assure that the Champenois, both vignerons and négociants, conform to rules regarding vine cultivation and the production of wine. Since 1945, vignerons and négociants have jointly monitored the industry through the Comité interprofessionnel du vin de Champagne (CIVC), which is controlled by the various professional groups within the industry. Ultimately, however, it is the state that assures compliance with laws regulating champagne production. The regulations and interprofessional cooperation that mark the industry today are the result of the struggles between vignerons, négociants, and the state in the late nineteenth and early twentieth centuries chronicled in this study.

The reader will by now have noted that there are a number of "champagnes." In French, it is fairly easy to distinguish between these variations:

"la Champagne" is the region, the old province; "Champagne *viticole*" is the vineyard within the region; *le champagne* is the wine; and the Champenois are the people of the region. Translating these into English creates a number of challenges. In order to avoid confusion, I use a lowercase "c" for the sparkling wine ("champagne") and a capital "C" for the region where the wine is produced ("Champagne"). When discussing the "vineyards of Champagne," I am referring primarily to the areas of vine cultivation within the department of the Marne. This is, of course, a highly controversial delimitation, as Chapters 5 and 6 demonstrate. But because most of the area devoted to vine cultivation falls within the Marne boundaries, this distinction is not without some justification. Likewise, when speaking of the Champenois, I refer to those in the department of the Marne who are in some way connected with viticulture, regardless of how tenuous that connection may be in the final analysis. As the following pages will confirm, however, determining what it means to be Champenois, to be connected with champagne, was much more difficult than the French words and their English translations might suggest.

CHAPTER Two

CONSUMING THE NATION

Champagne Marketing and
Bourgeois Rituals, 1789–1914

In *L'Accord fraternel*, a print published at the outset of the French Revolution, figures representing the three estates toast in a moment of revolutionary fraternity, each holding up a glass of wine. The commoner, complete with the celebrated tri-cornered hat, short jacket, clogs, and knee-length breeches, raises a simple goblet commonly associated with ordinary red wine. A member of the clergy, with flowing vestment, lifts the rounded, bulblike glass often used for consuming the wines of Burgundy. The aristocrat, wearing the breeches, ruffled shirt, and adornments of his station, lifts the unique fluted glass created to accentuate the sparkle of champagne.[1] In an era when symbolic expressions were often laden with political meaning, even ordinary objects (like wine glasses) and everyday customs (like offering a celebratory toast) were signs, conveying meanings that could be potential sources of political conflict (fig. 1).

This iconography paired champagne with abundance, celebration, and fraternity. Held in the hand of the aristocrat, champagne signified the opulence and privilege that came with birth in the social hierarchy of ancien régime Europe. Although the advent of the French Revolution brought the representatives of the three estates together in a moment of revolutionary fellowship, it had not eliminated social distinctions. Champagne, a rare, luxury item at the end of the eighteenth century, was reserved for the wealthy. Despite its affiliation with the fraternal gesture, champagne did not represent a social leveling, a radical impetus toward equality. A fluted glass and sparkling wine, much like a liturgical robe or a tricornered hat, was part of a complex language of signs that reproduced power structures. Wine symbolically recreated the social hierarchy, while the act of the celebratory toast produced an impression of social cohesion, a sense of restricted equality. Champagne appeared as part of symbolic forms that sought to mix old traditions and social hierarchies with new, revolutionary principles of the nation and rituals of national solidarity.

This complex association of champagne with social distinction and fraternal union evolved as both the center of power and national sentiments shifted in Europe's so-called bourgeois century. New techniques of producing, selling, and distributing champagne emerged in tandem with the advent of republican regimes and nation-states that, throughout the nineteenth century, were dependent on the opinions of ever-broader elements of the population. Champagne continued to appeal to the taste of elite clients who could afford this luxury product, while, at the same time, becoming part of the new, mass consumer culture that emerged after midcentury.

Almost a hundred years after the French Revolution, champagne became an "obligatory adjunct" to the social rituals of the emergent bourgeoisie of Europe. As one clever British observer noted in 1882, "We cannot open a railway, launch a vessel, inaugurate a public edifice, start a newspaper, entertain a distinguished foreigner, invite a leading politician to favour us with his views on things in general, celebrate an anniversary, or specially appeal on behalf of a benevolent institution without a banquet, and hence without the aid of Champagne."[2] Such remarks point to the centrality of champagne to bourgeois society during the long nineteenth century. Before World War I, champagne became a subject of mass culture, a centerpiece of bourgeois society invested with symbolic capital that paid enormous dividends to the wine-producing community of Champagne.

Regional producers of champagne, the champagne merchant-manufacturers, or négociants, adapted and shaped a mythic present and past for their commodity within these new frameworks. The years after midcentury, in particular, were ones of rapid change, when social groups and their environments were dramatically transformed, resulting in a search for "new social devices to ensure or express social cohesion and identity and to structure social relations."[3] Champagne was central to this process. It was used to delineate social boundaries and create solidarities; it became a common denominator for those claiming membership in various social groups; and it was integral to numerous nineteenth-century traditions and rituals. Whereas some of these traditions were consciously invented and constructed, others evolved more informally, with varying degrees of deliberate construction. Whether invented or evolved, these traditions, rituals, and images became a part of the business of selling champagne.

No single group, agency, or institution had the power to create or consolidate the rituals and images surrounding champagne. The négociants in the department of the Marne who controlled the small to medium-sized champagne firms were, however, in a unique position to profit from and to shape cultural attitudes about the wine and its consumption. Négociants were in the business of selling champagne, a business that, over the course of the century, evolved from dealing in "expensive wine aimed at a luxury market" and a discrete group of privileged clients to systematically promoting champagne and regional brand names among the burgeoning mass of middle-class consumers.[4] Before the sophisticated advertising and consumer surveys so ubiquitous in the twentieth century, the négociants of Champagne attempted to find a means to strike notes that would resonate with consumers. Although their marketing tactics may seem primitive by today's standards, the négociants experimented with advertising methods that are very much part of a modern marketing repertoire: packaging of the commodity, public relations spectacles, print journalism, and a promotional organization in the form of the Syndicat du commerce des vins de Champagne. The négociants had a "direct and intense need to understand and communicate effectively with their audiences" in order to sell their product to the relatively select customers who could afford the luxury commodity.[5]

Communicating with their customer base was a central preoccupation of négociants from the earliest days of the industry. For much of its wine-producing history, the Champagne region was known for its nonsparkling red wines sold in barrels by brokers. In the early eighteenth century, these still red wines were best consumed when young and were much sought after in the Paris market following the fall harvest each year.[6] Brokers dealing in both popular and elite wines who had strong connections with Paris wine merchants and consumers were best situated to respond to demand. Many of the families who dominated the regional wine trade were "hybrid, with a foot in the very lowest reaches of robe nobility and the highest ranks of the national wine trade."[7] Correspondence from the period before the French Revolution shows that these merchant families carefully cultivated clients among the "Parisian world of affairs and the elite provincial world of officials and nobility."[8] Understanding their clients was, thus, made less difficult by the fact that these early négociants held a class position and displayed cultural tastes that linked them closely to their audience.

On the eve of the French Revolution, the region's "markets dominated the economic life of the province and drove the creation of new kinds of wine."[9] Merchants in the towns of Reims and Épernay had come to control the regional wine trade. While merchants in Reims had a larger geography of customers than those in Épernay, both groups were slowly abandoning the Paris market for ordinary wines as competition led to declining profitability. By the end of the eighteenth century, trade in ordinary wines from the Marne collapsed.

Responding to changes in demand, merchants shifted their attention to offering a select inventory of fine wines. Elite consumers with reputations for refined taste increasingly sought wines that were aged and bottled before shipping, thus solving problems associated with barrel storage over long periods of time. Responding to their clients, Champagne wine merchants moved away from the Paris market and toward more exclusive, cosmopolitan markets in eastern and northern Europe. To assure supply, some merchants bought large vineyards and invested in bottling and storage facilities. What they lost in volume of sales, they more than recuperated by inflating prices.[10] Transforming themselves from mere brokers into merchant-manufacturers, the négociants of Champagne offered their fashionable clientele, not simply a fine wine, but an exclusive, magical elixir—a sparkling, bubbly white wine.

Once shunned as a trick by producers to cover the harsh tastes of bad wines, sparkling wine became fashionable with the trend-setting *grand monde*. Historians now generally agree that "sparkling wine first appeared among the consumers rather than the producers because it was a by-product of conservation techniques rather than the deliberate outcome of winemaking practice."[11] The bubbles occurred naturally when wine was made late in the year and the winter cold paralyzed the yeasts that normally turn grape sugars into alcohol. Warm spring weather reactivated the yeasts, and the fermentation began again, producing a carbonic gas that created a slight "sparkle" in ordinarily still wine. Innovative winemakers simultaneously experimented with improving the quality of the region's blended wines. The result was a pleasant-tasting, effervescent wine that became a passion in rich and elegant circles. While the sparkling wine was expensive and beyond the reach of the vast majority of Europeans, it had acquired enough of a popular reputation by 1789 to make it a powerful symbol in revolutionary literature.

Growing demand for sparkling wines of the region required that négociants build up inventories, improve wine quality, and market their product directly to clients. During the years that straddled the French Revolution, the champagne houses aggressively cultivated new clients through personal relationships. Itineraries for the typical house *voyageur* tell a story of constant travel to England, Prussia, Poland, Austria, Switzerland, Italy, Holland, and smaller states in between.[12] Undoubtedly, noble pedigrees, some acquired during the waning days of the ancien régime and others through loyalty and service to France's new emperor, Napoleon, along with connections to the emerging capitalist elite, facilitated access to customers.

The success of the earliest champagne families—such as the Ruinarts, Moëts, and Clicquots—have become the stuff of legend. Stories of the seduction of the European nobility by dashing négociants abound. Claude Moët is said to have started by convincing Madame de Pompadour, King Louis XV's mistress, that champagne was a necessity to any successful soirée. She, in turn, made it a court fashion by declaring it the only wine the consumption of which left women more beautiful. The founder of the Heidsieck house had numerous bold adventures in Moscow on a white stallion that won the admiration (and the regular business) of the tsar. The Ruinarts became regulars at the court of Charles X, where they were rewarded for their loyalty and regular supply of sparkling wine with the title of *vicomte*.[13]

Thomas Brennan has concluded that "many of the most prominent companies of modern Champagne owe their creation to the half century of growth spanning the Revolution."[14] In France, where fortunes changed quickly in the tumultuous years of revolution, representatives of families like the Moëts moved effortlessly from supplying ancien régime aristocrats to supplying the loyalists of, first, the Napoleonic empire, and, then, the Restoration.[15] While the Moëts quenched the thirst of France, the famous Veuve Clicquot opened new markets in Russia by sending a trusted representative there with a boatload of champagne in the aftermath of the

Napoleonic wars. Colorful stories describe how the irrepressible Widow Clicquot turned the French defeat into a marketing coup by throwing open her cellars to thirsty Russian troops. Seeing Russian officers freely polishing off bottle after bottle of her champagne, she reportedly exclaimed with a knowing smile, "Let them go at it! They're drinking? They'll pay!"[16] Over the century that followed, Russia became Clicquot champagne's top market, and its brand name became synonymous with sparkling wine there.[17] As the elite shifted their notion of power from a military to a cultural model, the Veuve would be attributed with the "peaceful conquest" of Russia for the industry.[18] French champagne triumphed where Napoleon's Grande Armée had failed.

No less impressive are stories of the brilliance of individual sales representatives selling champagne to the European bourgeoisie in the early years of the nineteenth century. One representative of the Clicquot firm, for example, attracted hundreds of new bourgeois clients in Königsberg with a single shipment—totaling over 10,000 bottles—in 1814.[19] Other firms in Reims and Épernay may have made less spectacular shipments but had equally striking sales figures. Their success can be measured by official statistics from the Reims Chamber of Commerce. Within forty years of the collapse of the Napoleonic regime, these firms were measuring their shipments in millions of bottles: by 1844, for example, approximately 2.2 million bottles of wine were sold in France annually and nearly 4.4 million more abroad.[20]

The deliberate creation of the sparkling wines of Champagne, Roger Dion observes, marked the "birth of large-scale commerce, new in the viticultural history of France."[21] Expanding markets were hampered by technical difficulties that plagued the industry in its infancy. Experiments with new processing techniques, discussed at length in Chapter 3, greatly accelerated production in the early nineteenth century, reducing losses and the overall price of sparkling wine. With these improvements in supply, and growing transportation and communication networks, sales no longer had to be limited to face-to-face communications or the bravado of dashing négociants. Moreover, clients were no longer limited to the wealthy circles of court society. As the ranks of the bourgeoisie expanded throughout Europe and North America, so, too, did the pool of consumers capable of purchasing luxury wines from Champagne.

Champagne marketing in these shifting circumstances meant adapting to a new customer base and developing new sales strategies. Wine marketing was dependent on cultural factors that could limit the manipulative powers of manufacturers. One challenge faced by champagne négociants was appealing to an audience of diverse consumers in Europe and North America. Historians have highlighted the diversity within the ranks of the bourgeoisie and the varieties of bourgeois experience in the nineteenth century. Indeed, the contours of the bourgeoisie shifted over the course of the century. By the later part of the nineteenth century, ownership of appropriate goods and participation in rituals of solidarity became important for defining class boundaries and membership. Although this diverse group of bourgeois consumers ultimately "sported many cultural styles," there was also a certain coherence of response.[22] This was the response that the négociants hoped to evoke in new promotional efforts by associating champagne with existing or evolving forms of sociability.

Even with this expansion of potential consumers, the importance of the client relationship remained central to champagne marketing. Branding strategies—creating brand-name identifications—was one method that négociants used to build a client relationship across great distances. Champagne "brands"—simply the family name of the founder of the firm—were used throughout the early part of the century to identify the origin of bottled wines from the region. Rarely, if ever, did the appellation "champagne" appear on a bottle; in keeping with the personalized nature of the market relationship, consumers looked for the family brand as a guarantee of quality.

By mid-century, brands appeared, in simple form, on labels attached to bottles delivered to individual clients. As négociants ceased to court the client directly, the family name on the label took on added importance as a form of personal assurance as to the quality and uniqueness of the product within the distinctive bottle. The price of champagne, the U.S. consul in Reims reported, depended, "principally on the reputation of its manufacturer; wine with the marks and labels of a well-known or celebrated maker sells for double the price of the same wine with an unknown brand."[23] To generate this brand-name identification, and therefore expand into new markets, particularly abroad, financial resources, personnel, and specialized skills were required. Robert Tomes, a contemporary observer, noted, "Strenuous efforts were made to give them [brand names] circulation."[24] Such strenuous efforts proved successful: of the 28 million bottles of champagne sold by 1900, for example, almost 23 million were sold by established brand-name producers.

Sales of sparkling wines soared after mid-century: 5.9 million bottles were sold in 1850; 8.1 million bottles in 1870; 20.4 million bottles in 1880; 25.7 million in 1890; and 28 million in 1900. The staggering wealth of the main Champagne families and the legendary client relationships of the industry's early years contributed to the enshrining of négociants as the heroes of the industry by admiring contemporaries. Négociants did little to discourage this admiration and, in some cases, actively promoted the mystique surrounding themselves and their firms, much as they promoted sparkling wine itself. These promotional efforts often took the form of official biographies of the founders of individual firms. The British writer Charles Tovey, who went to Champagne to conduct research for a book on the region in the 1860s, reviewed a collection of these biographies and concluded that the négociant families were uniformly portrayed to the public as "a superior race, heroes, or something more, celebrated in song and immortalized in history."[25] Tovey highlighted what other commentators also noted: the ongoing process of creating a mythic present and past for champagne brands.

Creating a "mythic" past that linked the wine and a family name to tradition and honor could add respectability and status to a brand. Given the importance of notions of honor and respectability connected to family within bourgeois society of the nineteenth century, it is perhaps not surprising to find the négociants concerned with infusing their brand names with a distinct *héritage*, as the historian Adeline Daumard has termed it—a heritage, legacy, or patrimony.[26] One means for the négociants to achieve this was to flaunt their aristocratic titles or, if they could not claim a pedigree from the earlier part of the century, to seek to obtain one through marital alliances with members of the French nobility. Entry into the aristocracy came mainly via marriages between the daughters of négociants and the sons of aristocratic families. Such was the case of the Cliquots, for example, who married their only daughter to the comte de Chevigné and gave the couple a magnificent château overlooking the family vineyards near Boursault.[27] Marital alliances provided négociants and their families with noble titles and links with a venerable past. Increasingly stripped of their former pejorative meaning, noble titles could be adapted by the bourgeoisie, the "new aristocracy, the nobility of the nineteenth century."[28]

Noble titles served to link wealth and authority in the new bourgeois civil society with the "remaining trappings of honor," in this case, aristocratic standards of consumption, family name, and personal reputation.[29] As Norbert Elias has illustrated, the nobility were specialists in elaborating and molding social conduct, which was, in the course of diffusion to other social strata, modified into a lasting cultural legacy.[30] Champagne and the reputable "noble" families whose names appeared as brands on the bottle were linked to "honorable" consumption and tradition, which could bridge the gap between the middle-class ethos of frugality, self-denial, and civic responsibility and the new consumer culture of the late nineteenth century.

For the négociant families, who willingly exchanged their personal wealth for status, obtaining noble titles was a shrewd marketing strategy. Without the vestiges of feudalism, titles and family coats of arms still held a luster of honor and tradition. Both the titles and the symbols of nobility could be added to the firm's name or logo, giving a certain air of distinction and a connection with a preindustrial tradition that could be more comforting than the dizzying reality of the industrial world. By the 1880s, champagne wine labels often featured brand names printed in bold, gold lettering flanked by symbols of royalty such as crowns or lions or accompanied by the family coat of arms. Indeed, gold was the most consistently used color on wine labels in the late nineteenth century.[31] Along with gold lettering and coats of arms, the use of words associated with nobility—"royal," "marquis," "prince"—and royal images—a double-headed eagle (associated with Russia's Romanoff dynasty) or lions (associated with kings and royalty)—gave even the less-expensive sparkling wines that appeared on the domestic market an air of honor, luxury, and timeless tradition. The firm of Jules Félix Fournier, for example, used a depiction of the Russian royal eagle combined with the brand name "Romanoff" to sell its wine. For the Venoge firm in the same year, the "noble" status of the wine is depicted through the use of a majestic gold lion, presented along with the noble titles of the Venoge family.[32] This connection with social exclusivity and status, Leo Loubère observes, constituted a "form of snob appeal," which became the hallmark of champagne négociants' advertising.[33]

Négociants and their agents went beyond these displays, however, continuing their legendary maneuvering to have their champagnes literally in the hands—and stomachs—of the rich and famous. Firms jockeyed to become *fournisseurs brevetés* or *privilégiés* (purveyors by appointment) to the royal courts of Europe. On their labels, négociants proudly proclaimed the connections between their firms and royal patrons such as the king of Spain.[34] More exotic royal courts could also have advertising appeal: dispensing with the typical symbol or brand name of the firm, one label announced in bold letters that this champagne had been chosen for a celebration by the Khedive of Egypt.[35] Champagne was linked to an "upscale"—to use modern advertising jargon—clientele, playing on the desire of bourgeois consumers to distance themselves from the conditions and values of the popular classes. Jean-Paul Aron notes that after the 1880s, the petite bourgeoisie were anxious to detach themselves from the proletariat.[36] Gastronomy and fine wines, like champagne, were one means of symbolically distancing themselves from the working classes. In creating this distance, négociants did not invoke the middle-class ethos of frugality and self-denial but rather appealed, by association with the charisma (defined by Max Weber as the possession of "supernatural, superhuman, or a least specifically exceptional powers or qualities")[37] of celebrated imbibers of champagne, to bourgeois aspirations to transcend the mundane, day-to-day social realities of what was sometimes called the "bourgeois century."

Association with royalty, nobility, and celebrities was not unique to champagne advertising. As early as 1834, an advertisement in the *Edinburgh Review* for "Mr. Cockle's Antibilious Pills" boasted recommendations from "ten dukes, five marquesses, seventeen earls, eight viscounts, sixteen lords, one archbishop, fifteen bishops, the adjutant-general, and the advocate-general."[38] These notables who promoted Mr. Cockle's Antibilious Pills, however, were relatively obscure before the advertising campaign, lacking the charismatic draw of a true "celebrity." What was different in the case of champagne advertising was that the négociants were able to appropriate the acknowledged social status of known "celebrities" for their commodity. These had symbolic value by virtue of their connection to what Clifford Geertz has called "the active centers of the social order."[39] When champagne négociants exploited the images of public personalities, however, they actually enhanced the prestige of those celebrities simultaneously with that of their glamorous product, and champagne's ability to create charisma could then be marketed to the bourgeoisie.

The construction of links to charisma and aristocratic standards of consumption were not exclusively the result of the commercial maneuvers of négociants in Reims and Épernay. Improvements in transport, new processing techniques, and reduced production costs made it possible to market wines and other formerly local specialties of France to an ever-broader group of consumers both at home and abroad. Some of these products—such as roquefort, cognac, brie, and champagne—developed reputations for excellence that extended far beyond their immediate production area. Taste professionals touted these products as essential to the art of eating and drinking, making them a part of the culture of ingestion, a tradition of superior quality and savoir vivre that was often dated back to the Roman era. Champagne's privileged place within the pantheon of French gastronomic quality, as we have seen, however, had a much less distant origin.

The négociants as a group were also relatively new to regional bourgeois society. In Reims and Épernay, much as in Rouen and Lyons, there was a certain antipathy toward "newcomers," whose ranks were swelling as the regional wine market prospered.[40] Marital alliances between members of champagne families and those of the commercial elite, mainly centered in textiles, were rare. The few prominent exceptions, like the Veuve Clicquot–Ponsardin marriage and the Pommery-Greno marriage in the early nineteenth century, gave rise to the notion that an "interlocking directorate" of wine and wool dominated the region. Generalizations about an "interlocking directorate" are, however, based on the marriage patterns of wine merchants in Bordeaux. The Bordelais merchants did tend to marry into local commercial circles, creating a number of interconnected families between industries.[41] Champagne manufacturing, in contrast, was a latecomer to the regional economy and society, where champagne négociants, both large and small, were viewed as nouveaux riches.

The proliferation of champagne firms and the meteoric financial success of their négociants were often viewed with suspicion. There was a long history of tension between the champagne négociants and other commercial elites, particularly their counterparts in textiles. Charles Tovey observed with distaste that "the accumulations of enormous profits, are evidenced by the palatial residences, as well as the large possessions belonging to the [champagne] magnates of Rheims and Épernay." Fortunes were built on reputation, but Tovey noted that the methods of obtaining this reputation were for many in the region suspect:

advertisements cunningly worded, extra allowance to wine merchants who will promote the sale, bribes to hotel-keepers and proprietors of steam-boats, the same to the managers of public establishments, paragraphs in newspapers that at such a dinner the Champagne was So-and-so's, and was pronounced to be of extraordinary quality; fees to waiters at hotels, and gratuities to stewards and butlers in the service of the nobility. . . . All these manoeuvres are followed up by a well-organized method of touting. Showy labels meet you at railway stations; everything is done that is possible to familiarize the public with the name; and all these combined attractions have successfully brought many a wine into a demand that its real quality did not deserve.[42]

Given the evolution of French society and the cultural constraints on entrepreneurship, the enormous profits of the négociants and their propensity toward "showmanship" did not correspond to the correct social function demanded of entrepreneurs in France.[43]

Noble titles were one way to resolve this dilemma. Association with aristocratic tradition could provide a mechanism for achieving at least some level of social approval within the region, perhaps providing contemporaries with a means for dismissing négociants' inappropriate behavior, since being "noble" had not lost all of its former trappings of ostentatious displays. Moreover, noble titles could mask the négociants' public pursuit of economic gain. Money alone was not sufficient to bring them honor and status. Some snickered that "the best blood of France can always be purchased by the heaviest purse," but, by restructuring their past, the négociants had to be dealt with as part of the existing social order, not as mere parvenus.[44]

Marriage into a champagne industry family in many ways provided an ideal source of income for nobles. William Reddy discusses the problem of seeking both profit and honor in French society in the nineteenth century in an analysis of Balzac's novel *Père Goriot*, in which Eugène de Rastignac, a young nobleman, wishes "to possess luxuries or gold not for their own sake, but because he could not gain honor except by appearing already to possess the increased honor he sought. This is why it was so important to appear already rich, to act as if one was already familiar with the peculiar customs and expectations of the *beau monde parisien*." With the new laissez-faire principles of the postrevolutionary social order, wealth and honor were tentatively linked: money could bring honor to those who possessed it, but, sometimes in order to obtain wealth, individuals were forced to dishonor themselves.[45]

The need to demonstrate an old French lineage was complicated by the immigrant origins of some of the most prominent négociant families. Non-Gallic names were conspicuous among some of the region's prestigious brands. The frequency of non-Gallic names among the *grandes marques* in particular drew considerable attention from industry supporters and critics alike. "There is, in fact, not a single wine establishment in all Champagne which is not under the control, more or less, of a native of Germany," Robert Tomes observed.[46] Names such as Koch, Mumm, Giesler, Heidsieck, Kunkelmann, and Walbaum point to the sizable number of foreigners, particularly Germans, who established themselves in the champagne industry during the nineteenth century. Although most immigrants were German, there were also a number from the Low Countries, Switzerland, and even Britain.[47] With the largest markets for champagne before 1850 found in the German states, it was not surprising that a number of established firms recruited German-born apprentices or German-speaking sales staff. These foreign apprentices were in an ideal position to establish their own firms, knowing the industry and possessing both the necessary language skills and family networks abroad. For French merchants, however, there seemed something slightly distasteful about those who profited after being initiated into the secrets of champagne making. One French-born négociant grumbled in 1845 that "the unhappy but nevertheless very real French unwillingness to study foreign languages [meant that they had to employ] young German clerks, to whom was confided the care of foreign correspondence. Many of these young persons, who thus found themselves initiated into the secrets of champagne making, well knew, with an intelligence that we are the first to acknowledge, how to profit from the exceptional opportunities they were offered and set up their own businesses."[48]

In a relatively new industry that had few winemaking traditions, advances or innovations in wine production or blending techniques were closely guarded secrets. The establishment of independent firms by these "foreigners"—foreign to the region and to the industry—many of them former clerks or sales staff, rankled some of their competitors. Moreover, in a period when wine and festivity were increasingly associated with a uniquely Gallic inheritance, the "foreignness" of the creators of this French drink was hardly an asset. Aristocratic links masked this foreignness, linking the négociants with Gallic traditions and integrating them into the social fabric of the region. Indeed, the archives provide numerous examples of nobles serving as "silent partners" in champagne manufacturing.[49]

The career of the négociant Mathieu-Édouard Werlé is instructive in highlighting the power of creating a Gallic *héritage*. Werlé, an active participant in the political and social circles of Reims for much of the nineteenth century, is often a central character in industry narratives. His career began in August 1821, when he presented himself at the offices of the Maison Veuve Clicquot–Ponsardin in Reims. He requested a meeting with the formidable Veuve Clicquot, the legendary widow who, since the death of her husband in 1805, had transformed her family's modest wine house into one of the leading champagne producers in Europe. This bold, self-assured twenty-year-old, the story continues, had traveled there from the town of Wetzler (in the principality of Hesse) and was determined to obtain an apprenticeship at the Reims office of the famous widow, already well known for launching the careers of several other German immigrants. Werlé impressed Clicquot with his intelligence and his resolve, and she hired him immediately.[50]

By 1831, ten years after knocking on the door of the Maison Veuve Clicquot–Ponsardin, Werlé had advanced from his position as clerk to become a partner in the firm. Five years later, he married into one of the oldest woolen-manufacturing families in the Marne (which fortuitously had close ties to another wine house, Maison Roederer), thus strengthening his economic and social position in Reims and making him a French citizen. The politically ambitious Werlé capitalized on his situation to expand his influence. He became a judge in the Tribunal de commerce in 1838 and subsequently served as president of the court from 1846 to 1850.[51]

His rise to political power was as spectacular as his ascendancy at the Veuve Clicquot–Ponsardin wine firm. As one of Reims's leading conservatives, Werlé became a spokesman for the Second Empire.[52] His loyalty brought him many political appointments, and eventually he became "the most important political figure in Reims."[53] His alliances were solidified, and his firm gained an important economic boost through the marriages of his two children to wealthy, aristocratic heirs of families connected with the Bonapartist regime. By the time of his death in 1884, Werlé was one of Champagne's leading citizens; his family had a legitimate noble pedigree and Gallic *héritage*; his firm of Veuve Clicquot–Ponsardin dominated foreign markets and was among the most successful champagne houses in the region.

Immigrants appear to have settled permanently in the region, and by World War I, only a few had failed to take up French citizenship. In the case of G. H. Mumm, for example, failure to become French led to confiscation of the family's firm and property at the outbreak of World War I. Although many négociants were suspect on the eve of World War I, those who had been naturalized or married into a French family were cleared.[54] Evidence suggests that suspicions about immigrant négociants began to wane in the late nineteenth century as champagne became the common denominator that linked the négociants as a group within the region and their status as foreigners was sufficiently masked. United by shared economic concerns regarding the wine and its production, négociants focused attention, not on the distinctions and inequities between individuals and firms, but on the relationship of champagne—the wine, the region, and its capitalist, wine-producing enterprises—to the outside world.

❦

With sales of wine lagging because of the global economic downturn after 1870, a small group of firms came together in 1882 to create a powerful, unified négociant promotional organization. Between 1882 and 1900, the Syndicat du commerce des vins de Champagne (now called the Syndicat des grandes marques de Champagne) never included more then sixty firms, all of them adopting the title of *grande marque*.[55] If there was any lingering resentment within négociant circles toward foreign-born négociants, it was not apparent in the organization of the Syndicat du commerce. Nearly one-third of the membership in 1895 were first and second-generation immigrants.

Integration of négociant families into the social fabric also meant recreating "traditions" that reinforced hierarchical relations within the wine community. This is evident in the industry promotional organization. While the Syndicat du commerce was theoretically open to all "négociants or former négociants of wines from the department of the Marne," those who considered themselves *grandes marques* did not invite all producers to participate. There were no regulations or written procedures regarding the use of the term *grande marque*. Evolving market practices developed almost imperceptibly into "traditions," in this case defined by the Syndicat du commerce itself. According to industry literature, the grandes marques were known around the world because of their "respect for traditional rules of production." Respect for tradition and "une grande notoriété mondiale" seem to be the defining criteria for a grande marque.[56] It is interesting to note, however, that the title of *grande marque* appears to have been constructed in ways that are remarkably similar to the construction of the concept of the *grande bourgeoisie;* both were based on a combination of fortune, family, and tradition. By the late nineteenth century, the grande bourgeoisie closely resembled the former nobility: its members occupied or had occupied a place in the bourgeois hierarchy for generations, had important ties to the state, and displayed a preference for a certain lavishness and tradition that recalled the "aristocracy."[57] Similarly, the group of grande marque firms occupied an important economic position within the regional wine industry for several generations, had long-established ties to the state, and were linked to aristocratic tradition. The hierarchy of firms was structured much like the bourgeois hierarchy of France.

Tradition among the négociants and their firms was increasingly defined by the grandes marques–dominated Syndicat du commerce. Calling on the regional héritage, the organization insisted that all wines bearing the name "champagne" be made from grapes harvested in a broad territory loosely defined as Champagne and that all aspects of their handling (*manutention*) be completed within the region. Pamphlets distributed by the Syndicat du commerce highlighted the benefits of this "tradition," creating an image of harmonious class relations in which all social groups shared in the general prosperity of the wine industry. Most important, these pamphlets stressed the "authenticity" of the products of the region created according to traditional methods. In an age when the public was increasingly focused on the "quality" and purity of wine, this reference to the authenticity of champagne carried a certain resonance.[58] Authenticity was important, particularly in foreign markets where competitors increasingly employed the label "champagne" for nonregional sparkling wines. Regional producers focused on promoting authentic champagne outside France, where the consumer was "aware of the magic" of the "true wine of champagne."[59]

The emphasis on foreign markets reflected the concerns of the core of the Syndicat du commerce—the producers and firms that were among the largest exporters of champagne—between 1882 and 1914. It should be noted, however, that numerous firms with relatively low levels of foreign trade were also included as Syndicat members. This was the case with the firm of Charles de Cazanove in Avize, for example, whose sales were very modest in comparison with those of Deutz & Geldermann or Veuve Clicquot–Ponsardin. The Syndicat du commerce used this "mixed" representation to its advantage in the early twentieth century, when it was accused of creating a monopoly.[60] Despite, or perhaps because of, its rather exclusive membership, the Syndicat created the impression of a powerful, unified négociant presence. With the considerable economic power and national and international recognition of its membership, it was generally acknowledged both by those in the industry and by the French state as the representative body of the champagne houses and négociants. Under its auspices, a collective voice for négociants was discernible in public debates. It worked not only to protect the name of champagne abroad but also to modify domestic laws and regulations concerning octrois,[61] transport costs, and taxes for the benefit of all manufacturers, regardless of size. The results of its lobbying efforts would be shared, even if membership in the Syndicat was not.

When the Syndicat du commerce was formed in 1882, there was little reason for more modest firms to oppose its exclusivity or efforts. It gave the industry as a whole a powerful means of promoting champagne and deflected any bad publicity directed at négociants. The Syndicat not only kept a file of articles considered detrimental to the industry but often responded to them, officially denouncing the "misinformation" they embodied.[62] It argued that champagne was a unique beverage, connected with age-old traditions originating in a distinct region within France. Publicity pamphlets stressed that only sparkling wine produced within the confines of the Marne by regional producers should be considered "champagne." This attempt to promote the product category by linking the denomination "champagne" with a region and a group of producers was one of the key goals of the new organization. This is abundantly clear in the pages of the annual accounts of the work of the Syndicat du commerce. Much of the annual discussions are taken up with issues of protecting or reinforcing this link between the wine and the firms of Champagne through domestic and international legislation.[63]

An important bond was stressed between the wine (champagne), the region (Champagne), and the people (Champenois). This was particularly important as individual négociants pursued a number of long and costly legal battles in the French courts to prevent "champagne" from being used as a generic label for sparkling wines unconnected to their region. In a series of decisions that began in the 1880s and spanned two decades, French courts determined that "champagne" did not fall into the public domain as a generic processing technique. Although other sparkling wine was also the result of a second fermentation technique that produced gaseous bubbles in still wines, the courts ruled, it was ultimately the wine of the Champagne region that gave champagne its unique character. The name "champagne" specifically designated a natural product of the Champagne region that was rendered "sparkling" (*mousseux*) in a second processing phase. It was not a generic term for the second processing technique.[64]

By the late nineteenth century, these legal decisions were seen as only a partial victory. The courts had no authority outside of France, where many of the violations occurred. Producers outside of the region frequently used the name and often adopted the style of champagne labels and packaging.[65] "The wine of Champagne has suffered the fate of all great discoveries," wrote one négociant, "a multitude of imitations rushed forward, and the leprosy of counterfeits clung to its [champagne's] vogue."[66] The Syndicat du commerce engaged in a legal and legislative battle to protect the regional appellation from being used as a generic label for sparkling wines and the brand names of regional producers from imitation or counterfeiting at home and abroad. Under the auspices of the Syndicat, the négociants argued that the government needed to take action to protect the name "champagne" as the property of the French nation, guarded by the traditions practiced by the community of producers within the confines of the Marne.[67] The négociants were singling out champagne for special status among the wines of France, which were already "encrusted with national myths about the glory and genius of the French race."[68]

The active defense of the denomination "champagne" by the négociants and the Syndicat du commerce corresponded to an interesting shift in product marketing. Until the 1860s, the name "champagne" was rarely used on wine labels to identify the sparkling wines produced in the region. Négociants did not give the word "champagne" prominence on their wine labels, preferring the name of a prestigious vineyard, or *cru*—such as Aÿ, Sillery, Verzenay, or Bouzy—followed by the words *mousseux* or *grand-mousseux*.[69] This is similar to practices in other wine regions where the names of châteaux or villages (for example, Châteaux Margaux or Saint-Julien) were connected with a famous cru. Given the complex blending process involved in champagne production, however, it was difficult to link the final product to any particular vineyard. Moreover, with the demise of regional still wine markets, the prestige of a cru increasingly had little meaning to a larger audience of consumers.[70] "Champagne," followed by "mousseux," occasionally appeared in place of names like Aÿ and Sillery on labels in the late 1860s.[71] By the late 1870s, there was a gradual increase in the use of "champagne" on labels.[72] And, with the appearance of the Syndicat du commerce in the 1880s and new promotional efforts, "champagne" appeared with the name of the location of the manufacturer rather than the cru identification.[73]

For the négociants who used the regional appellation, "champagne" represented not simply a wine but, more important, a community of producers that contributed to the glory of the French nation. Pamphlets and articles by the Syndicat du commerce reinforced this belief by popularizizing the unique genius of both the regional wines and the community by means of a number of myths about the industry and its "founders." Probably the most influential and most enduring concerned the "discovery" of sparkling wine. The largely forgotten monk Dom Pierre Pérignon was resurrected as the inventor of sparkling wine and the founder of the champagne industry. The creation of the myth is generally attributed to Dom Grossard, a former Benedictine monk at the Hautvillers Abbey of Saint-Pierre. Reduced to a simple parish priest after the French Revolution, Grossard was determined to memorialize the great "achievements" of the abbey. He began his campaign in 1821 with a letter to the deputy mayor of the town of Aÿ, informing him that a local monk had invented champagne. "As you know Monsieur," he wrote, "it was the celebrated Dom Pérignon . . . who found the secret of making white sparkling wine."[74] In his missionary efforts for public recognition of Pérignon, Grossard attributed virtually all the advances in Champagne viniculture and viticulture to him.

Although there was no evidence to support Grossard's assertions, or the various versions that were subsequently transmitted, the story of the "miraculous" accomplishments of this monk spread over the course of the century.[75] By the 1860s, authors writing about the region unquestioningly attributed the discovery of the "secret" of sparkling wine to Dom Pérignon, who by now had been transformed into a blind monk who used his highly developed senses of smell and taste to create the finest blend of wines.[76] But Dom Pérignon's "resurrection" did not have widespread recognition until the Syndicat du commerce began to use it for commercial purposes. At the 1889 Exposition universelle in Paris, the Syndicat provided the public with an illustrated pamphlet that reproduced the Dom Pérignon story, declaring him to have been the "father" of sparkling wine.[77] After tasting the wines of Champagne, the visitors could carry away this souvenir, sharing the Dom Pérignon story with friends and family in France and abroad. In 1896, the Syndicat produced a pamphlet entitled *Le Vin de Champagne* that unequivocally stated that Dom Pérignon had "discovered" champagne by following "ancient traditions."[78]

International visitors to the Champagne region, familiar with the story of Dom Pérignon, often expressed surprise that there were no memorials erected to the founder of champagne.[79] Indeed, it would not be until 1910, a year before the bloody revolt in the region, that a statue commemorating Dom Pérignon was erected in Épernay.[80] The American consul was stunned when he went in search of the famous abbey and local vignerons explained that it had been "justly" destroyed by revolutionaries.[81] If there were no memorials to Dom Pérignon within the region, however, it did not take away from the drinking public's faith in the Dom Pérignon story. By the beginning of the twentieth century, the legend was established as fact. Visitors to the Brussels Exposition universelle in 1910 could visit an "authentic" reproduction of the "abbaye royale d'Hautvillers," complete with presses and barrels allegedly used by the famous monk.[82] The success of the legend, however, was best exemplified in the widely circulated *Petit Journal*. In June 1914, the bicentennial of Dom Pérignon's "discovery" was commemorated with a special color-illustrated issue. Under a drawing of Dom Pérignon, looking strikingly like St. Francis of Assisi opening a bottle of champagne, was the caption: "It was exactly two hundred years ago that Dom Pérignon, a Benedictine monk, discovered the art of making the wines of Champagne sparkle."[83]

If Grossard's efforts to promote the Dom Pérignon story were motivated by a desire to reestablish the prestige of the Catholic Church in France, the myth also served the more secular purpose of seducing the champagne-drinking public. A wine originally associated with dissipation and hedonism was now believed to have been invented by a monk; the holy origins of champagne helped to legitimize a drink originally associated exclusively with aristocratic frivolity and decadence. The story of a simple monk "discovering" bubbles in the region's fine wines encouraged the impression that champagne made within the confines of the Marne by regional négociants was *the* original sparkling wine; others were not authentic, being simply imitations, which the informed consumer should not accept. The legend supported the claims of the champagne industry to be one of the glories of the French nation—and hence worthy of government protection from the challenge of competing sparkling wine producers both at home and abroad in the late nineteenth century.

Despite the almost industrial techniques used in sparkling wine production, the Dom Pérignon myth distanced champagne from any association with assembly lines, technology, and backbreaking labor. The monk's "simple" invention was cultivated in public relations campaigns to create an image of champagne as being as effortless to create as it was to drink, a symbol of a balance between old-world traditions and the "good life" of the modern period.

Champagne and wine brand names offered consumers a sense of continuity. In the face of the complexities of the new social world of the late nineteenth century that made the individual consumer feel incompetent or insecure, the timeless tradition of champagne offered reassurance that one was upholding the highest standards of social intercourse, thus reinforcing the individual's sense of membership in a civilized community. This is most apparent from the labels that adorned champagne bottles from the period. What is striking about these labels is that, unlike modern ones, they tend to include elaborate pictures and written texts, signs meant both to instruct and to appeal to the individual client. A public register was created in 1859 to record the trademarks and labels used by the champagne négociants and their firms. Although firms were not required to register their labels, many took advantage of the recording mechanism, which provided them with a legal basis for pursuing counterfeiting cases in court. Hundreds of labels, now carefully preserved in the Archives départementales de la Marne in Châlons-en-Champagne, were entered into the register between 1859 and 1902. This record, combined with surviving promotional pamphlets and posters, provides a substantial database for analyzing marketing trends.

Wine labels, posters, and pamphlets, of course, present many of the same interpretive problems as conventional texts. Historians cannot determine how the public received these advertisements, but, as Roland Marchand has noted in his study of American advertising, "neither can we prove the effects of religious tracts, social manifestos, commemorative addresses, and political campaign speeches on their audiences."[84] The goal of the champagne négociants was to sell their wines; they would not have chosen or continued using messages that failed to achieve this goal. Thus, the choice of soothing words such as "pure," "special," and "love," which frequently appeared on labels, must have had substantial appeal, perhaps providing a certain nurturing reassurance and sense of belonging that the négociants, themselves, had long sought (fig. 3).[85]

Champagne could be used as a sign of class membership in public space. The emergence of restaurants and public banquets, bringing formerly private rituals into the public sphere, shifted gastronomy, including fine-wine consumption, to a central place in social life.[86] Consumption was a status symbol, and material goods were a visible symbol of rank. As Jean-Paul Aron has stated, it was "*à table* that the nineteenth century began to define itself; it is *à table* that business deals are made, ambitions declared, marriages arranged," and, in this way, food, drink, and their consumption became a part of emerging nineteenth-century rituals.[87] Bars and restaurants became focal points for public consumption, and manufacturers launched new brands at fashionable restaurants, like Maxime's in Paris. The "temple of champagne" was the rue Royale in Paris, home of Maxime's. Created in 1891 by Maxime Gaillard, this restaurant became the model for chic restaurants at the turn of the century. Launching a new brand of champagne at Maxime's was a way for champagne manufacturers to associate their wine with pleasure, wealth, and fashionable society.[88] Honor and status, embodied in fine-wine consumption and champagne brand names, were displayed in more visible, public ceremonies at the turn of the century. Champagne was deemed the chosen wine of princes and gentlemen.

Champagne was meant to project a class image. This class image, however, did not necessarily reflect the consumer's socioeconomic status with any fidelity. The objective was to sell champagne, and this meant setting the wine apart from the mundane realities of daily life. Négociants attempted to associate their wines with scenes of higher status, placing the product in upscale settings, linking wealth with honor, reassuring consumers that they were part of a community that was removed from the conditions and values of the popular classes. By the end of the nineteenth century, one contemporary food and wine expert wrote, at all bourgeois social events, champagne was "not a wine, it is *the* wine."[89]

The appearance of champagne at the fashionable nightclubs that sprang up during the Belle Époque did not appear to tarnish champagne's respectability and use in upper- and middle-class rituals; paintings and posters from the Belle Époque attest to the importance of champagne in the new music halls of the era. Using the symbol of champagne in such a way caused conservative critics to lash out at its inclusion in the new culture of entertainment. "It is perhaps a mode," wrote one critic, "but it is surely a heresy; champagne at the *cafés de nuit* is an invention of the devil."[90]

Although the wine could lead to excesses and misfortunes, it did not appear to be associated with the baseness of ordinary drink or public drunkenness that were so offensive to a collective bourgeois morality. Common wines and alcohol were part of public debate about alcoholism and public morality throughout Europe, but neither middle-class social drinking and excesses nor champagne figured in these discussions.[91] Wine, particularly in France, was viewed, in the words of a contemporary, as "good for one's health and good for the nation."[92] The connection between wine and health is probably most vividly revealed in the language of the late nineteenth century. Fermented drinks—wine, cider, beer—

were not designated in the French language under the word "alcohol" (*alcool*) and, thus, were not associated with discussions of alcoholism (*alcoolisme*) or excessive alcoholic consumption. Wine was termed, even among temperance organizers, a *boisson hygiénique*, a healthy drink.[93] Commercial interests in Champagne bolstered the linguistic association through marketing and legislative efforts. Specialty wines were marketed as a form of hygiene, such as "Grand vin de santé" or "Vin de Champagne diabetique." Meanwhile, health professionals at the Grande Pharmacie on the boulevard Haussman in Paris sold a house brand of "Médicinal Champagne" for a range of ailments.[94] Public drinking in the modern music halls might have been associated with decadence and moral decline by some critics, but for many bourgeois consumers, taking their cues from the medical community, champagne remained linked to the vigorous, saintly Dom Pérignon.

Throughout the Belle Époque, the middle classes increasingly embraced public amusements and the consumption that often accompanied them, but "only after dressing them with respectability in the guise of civic donation or charity."[95] In this guise, champagne was incorporated into a wider variety of celebratory rituals (fig. 4). Négociants actively promoted the association of the sparkling wines of Champagne with social gatherings, expanding the images of the wine's functions. "Mythic historic events," as Erving Goffman called them, or junctures and turning points in life, were celebrated by the opening of a bottle of champagne, marking the event as above the mundane. Wine labels increasingly suggested that religious events—such as baptisms and marriages—were occasions for secular celebrations.[96] Indeed, champagne could be used to mark many of the main events of the life course: *fiancé champagne* for engaged couples, *champagne nuptial* for newlyweds, and *bebé champagne* for new parents (figs. 5 and 6).[97] If there were doubts about when to serve the wines, the labels provided examples. On the label "Bouquet of the Bride Champagne," for example, the happy couple are shown as recipients of a celebratory toast offered by the guests at an elegant wedding banquet.[98] Champagne became a central part of these social occasions. Newspaper advertisements, particularly near holidays such as Christmas and New Year, associated family gatherings with champagne.[99] One observer noted in 1881 that the increased use of champagne at festive gatherings was "a charming fashion that is beginning to be more common."[100]

Champagne was part of new "traditions"—like nuptial banquets and secular baptisms—some of which were decidedly "modern." At the turn of the century, transport became a symbol of progress, modernity, and luxury. "Christenings" were extended to the launching of ships or first flights of airplanes. One of the more famous christenings came in 1902. George Kessler, Moët & Chandon's agent in the United States, created an enormous stir in both the American and European press when he managed to substitute a bottle of his firm's champagne for a bottle of German sparkling wine at the highly publicized launching of the German emperor's new yacht, the *Meteor*, in New York.[101] For weeks following the event, the international press devoted front-page coverage to this famous christening, focusing attention on both the ritual and the symbolic importance of the wine.

Airplanes became one of the more important symbols of luxury and progress among the French bourgeoisie in the latter part of the Belle Époque. Adeline Daumard has noted that these symbols of luxury were so popular that bourgeois couples, who would never experience air flight, often arranged to have their photograph taken in front of airplanes.[102] Négociants worked to associate their wines with flight, creating aviation festivals and christening airplanes. Only a few years after the first flight, a French magazine cover featured a handsome couple on the wings of an airplane, toasting with champagne in the moonlight (fig. 7). The caption of the 1905 picture declared: "It is you Clicquot, Mumm, Roederer, Moët, and Pommery who triumph in the air! All the French *coqs* sing in the fields: the best motor is the wine of champagne."[103] By the eve of World War I, flight and champagne seemed intimately connected.

Champagne also became associated with leisure activities and sporting events. Labels featured pictures of horse races, sportsmen hunting and rowing, and healthy young people on bicycles (fig. 8). These advertisements suggested that social drinking, particularly drinking champagne, was an essential element of these leisure activities.[104] And champagne became a regular feature at "clubs" and at stylish outdoor events. Brilliant posters created by artists such as Pierre Bonnard and Leonetto Cappiello featured fashionable men and women, champagne glasses raised, celebrating the Belle Époque at clubs, picnics, or derbies. By 1900, the *Wine Trade Review* attested to the success of these efforts, declaring that champagne was "easily holding its ground as pre-eminently the popular wine for festive gatherings of all kinds."[105]

By the end of the nineteenth century, the champagne entrepreneurs had learned that one of the best ways to sell their commodity was to sell an ideology—an ideology of nation (from national identity embodied in monarchy to imperial expansion in Africa and Asia). Négociants responded to nationalist sentiment by creating wine labels that appealed to a sense of patriotism at home and abroad (fig. 9). Symbols of the nation—flags, battles, and soldiers in uniform—were central images on these labels. Special champagnes were produced and promoted as a way to mark contemporary political events, like the Franco-Russian alliance of 1893, or historic national events, and Columbus's "discovery" of America.[106] Firms promoted labels to mark the centennial of the French Revolution, featuring the famous Tennis Court Oath. One firm offered a different kind of revolutionary remembrance by creating a label that featured the name of Marie-Antoinette splashed across the tricolor flag.[107]

By the 1890s, wine labels began to reflect a general shift toward a more conservative nationalism. For the French market, one firm set out to appeal to anti-Dreyfusards during the Dreyfus Affair with the label of "Champagne Antijuif" (fig. 10).[108] Other firms appealed to French nationalists with images of Joan of Arc and words such as *Dieu* and *patrie* prominently displayed.[109] With European imperial expansion, champagne producers sought to capture the nationalist fervor of the scramble for overseas empire by creating commemorative labels such as "Champagne d'Orient" and "Champagne of India," featuring soldiers in exotic locations, or "Grand vin imperial," featuring two scantily clad "natives" holding large clubs. In an attempt to keep up with the patriotic fervor on the eve of World War I, one champagne firm simply modified the soldier and flag featured on its standard wine label—changing the color of the soldier's uniform and flag depending on the final destination of the wine.[110] Champagne had become a "resilient totem" able to support these "varied mythologies" about nations without having to account for contradictions.[111]

John Gillis has observed that in the rapidly changing environment of the late nineteenth century, there was "a profound sense of losing touch with the past."[112] No brand name or title could create a better sense of continuity with the past and a certain nurturing reassurance than the word *veuve*, or "widow," which appeared on the labels of all female négocants in Champagne. While men were "designated carriers of progress," women were seen as belonging more to the past as "keepers and embodiments of memory."[113] The "widow" as a genuine mother and a dedicated wife could provide consolation to those who feared becoming rootless in the new world of the late nineteenth century. Labels that appeared on champagne bottles after 1880 show a marked increase in the appearance of the title *veuve*. This did not mean that there was an increase in the number of firms headed by widows. For the most part, these new widows could best be described as "fantasy *veuves*," fictional widows from fictional champagne families. The firm of Mercier, for example, introduced the wines of Veuve Damas of Reims, a purely fictional widow, in 1885.[114] Mercier was one of the more innovative marketers within the region, with a well-established brand name. There are no records of what motivated the directors at Mercier to adopt a *veuve* label. But we can assume that the négociants at Mercier were attempting to create an advertising message that was more or less shared by their audience. Other male-owned firms followed the Mercier example, creating labels that featured the names of other fantasy *veuves*.[115]

In their capacity as wives and mothers, women frequently appeared on champagne wine labels. In 1887, the year of Queen Victoria's first Jubilee, champagne négociants attempted to use her charisma to promote their product.[116] Victoria provided upscale appeal, but she was commemorated mainly as a wife and mother. Her name and her image, in black mourning dress, appeared on labels of champagne bottles, much as they did on other commodities of the era. Other women appeared as allegories, such as Joan of Arc, a symbol of national identity.[117] The history of these "real women," much like the real *veuves* of the champagne industry, was systematically forgotten. As Gillis has noted, "Women and minorities often serve as symbols of a 'lost' past, nostalgically perceived and romantically constructed, but their actual lives are most readily forgotten."[118] The deeds of these women were not what gave them a marketing appeal. It was as a "genuine mother, attentive and devoted"—whether "mother" of a country or of a family—that these women acquired symbolic importance that could be exploited in the new commodity culture of the late nineteenth century.

Négociants attempted to increase the appeal of champagne to an ever-broader group or "mass" of consumers, male and female, young and old. Wines were, for example, produced and marketed for tourists at the Exposition universelle of 1878 in Paris, and for students celebrating academic success.[119] For these new consumers, the négociants introduced different qualities of sparkling wine. The wine house of Dufaut, for example, produced seven different quality wines: Grands vins de réserve, Carte blanche, Bouzy, Fleur de Sillery, Grand crémant, Royal Sillery, and Aÿ mousseux. Numerous wine houses used the term *carte*, associated with various colors—*or, blanche, verte, bleue, noire*—intended to denote the quality (and price) of the wine. Champagne firms with relatively unknown brand names tended to use terms like *royal, supérieur, spécial,* or *extra* along with *champagne* to give their wines added attraction. The least-expensive sparkling wines of the region were usually designated by the euphemism *fleur* or *petit champagne*.

With champagne's appeal to an ever-broader group of consumers by the turn of the century, a number of the grandes marques firms appeared to shift their marketing strategy, moving away from a presence in multiple markets to increasing market shares in countries where they already enjoyed brand-name recognition. Other négociants discovered that they could augment sales in the competitive foreign market by selling off-brands under the name of the importer. There are no accurate figures on the number of off-brands sold. With a certain lingering stigma attached to "cheap" champagne on overseas markets, few producers admitted creating these off-brands. Records indicate, however, that they were growing in popularity at the turn of the century. In the British market, for example, where nearly half of all exports were consumed in 1896, the Victoria Wine Company, one of the pioneers of off-brand champagne marketing, ran an advertisement listing twenty-three brand-name champagnes along with seven "house" champagnes at considerably lower prices.[120] Without the costs of distribution, marketing, and salaries for an overseas sales staff, these wines did not carry the elevated price tag of the more prestigious brands. An economist at the *Revue des deux mondes* estimated in 1894 that the average négociant made approximately 1 franc on each bottle sold for between 6 and 8 francs on foreign markets. It was, above all, he calculated, the costs of distribution that elevated prices.[121] Improvements in transport and the advent of large-scale distribution networks by 1911 lowered prices for off-brand champagnes, now generally selling for between 2 and 4 francs per bottle (compared with 8 to 12 francs for brand-name champagne in 1911), making them increasingly available for families of "modest means," particularly in France.[122] Although these off-brand wines were lower in quality, they were associated with the distributor or importer and did not jeopardize the reputation of the champagne house or the négociants that produced them.

With more sophisticated processing techniques, firms created new champagnes, such as "dry" and "extra dry," to appeal to the tastes of specific markets. By the exposition of 1889, for example, the Mumm firm featured three qualities of wine: Carte blanche (sweet); Extra-Dry (dry); and

Cordon rouge (very dry).[123] Dry wines were particularly popular in foreign markets and, despite being attributed to the wine-making genius of a single woman, these were generally associated with male consumers. Sweeter wines were generally popular in France, Germany, and Russia.

Sweet wines were deemed particularly appropriate and respectable for "ladies" to drink. Fashionable women in Britain often served champagne along with afternoon tea.[124] The success of this popular association is captured in a drawing by John Leech featured in *Punch* in the 1890s. A bourgeois family is gathered at the dinner table with glasses of champagne. The young son, returning from boarding school, is asked, "Now, George, my boy, there's a glass of Champagne for you. Don't get such stuff at school, eh?" The young man responds, "H'm! Awfully sweet. Very good sort for ladies. But I've arrived at a time of life when I confess I like my wine dry" (fig. 11). Symbolically, the young man was indicating his transition into the world of adult males through his rejection of the sweet champagne of his mother and the other ladies gathered around the table. Popular etiquette books in France went even further, forbidding the drinking of anything but the driest champagne at meals at which women were not present.[125] Men were instructed to envision a new bottle of champagne as a "beautiful woman" who after a voyage "needs to rest a little in order to present herself with all her advantages."[126] Centuries after Madame de Pompadour deemed it the appropriate wine for ladies, champagne remained a gendered beverage.

Négociants helped to fashion champagne as a consumer product not only through advertising and specialized products but through spectacle as well. Representatives and agents of champagne firms, individual négociants, and the Syndicat du commerce, with their publicity stunts and advertising efforts, acted as modern minstrels for champagne in the expanding marketplace. The bravado of champagne agents or the exploits of négociants at well-publicized events or international expositions generated enormous publicity for the industry, creating sensational news items and great "theater." For the Exposition universelle of 1889, for example, Eugène Mercier commissioned the largest barrel in the world, decorated with elaborate sculptures by the artist Navlet. Containing nearly 200,000 bottles of wine, the transportation of this barrel to Paris by twenty-four white oxen and eighteen horses received coverage in newspapers from Hungary to San Francisco. For three weeks, press reports focused on the progress of the barrel, keeping "champagne" and "Mercier" in the public eye.[127] Even those who could not afford champagne could delight in these stories and participate through name identification and popular imagery. By the beginning of the twentieth century, "simply the names of these wines became household words."[128]

Although négociants marketed to a broader group of consumers, there was no suggestion in advertising or promotional materials that the regional sparkling wines were intended to be "democratic," consumed by a clientele from a broad range of economic and social classes (figs. 12 and 13). Champagne represented "restricted equality"—the desire or demand for it might be democratic, appealing to the mass of consumers, but those who could afford to participate in its rituals and consumption were limited to an exclusive group of clients. As one observer noted, "the respectable négociant demonstrates that it is impossible to sell the real products of the Champagne vineyards at a low price."[129]

While intellectuals like economist Paul Leroy-Beaulieu reminded the elite on the eve of World War I that it "was not a crime" to drink cheap champagne, particularly at *banquets democratiques* or family gatherings for those of "modest means," no one should pretend that these wines were anything but *faux champagne*. There was an assumption that cheap sparkling wine was not authentic, since "it could hardly claim to be made with grapes from Reims or Épernay."[130] This served as ample justification for dismissing the popular consumption of champagne in cabarets. It was a given that authentic champagne had to be expensive, setting apart the wine and those who consumed it from less-affluent consumers of common wines.

On the eve of World War I, champagne was an essential part of social life among the European bourgeoisie. The sparkling wines of Champagne were part of a repertoire of symbolic devices that were used to delineate social boundaries, in France and abroad, distancing the bourgeoisie from the popular classes. In the changing world of the late nineteenth century, brand names and material goods denoting social status became particularly important for creating a sense of group identity. Commodities such as champagne could be employed in new rituals and "traditions" of membership that took the place of routines, customs, and structures that were increasingly obsolete or irrelevant. During this evolution, the "old"— whether noble titles or aristocratic standards of consumption—could be refashioned and reinvented to create something "modern" that offered a sense of continuity and a reassuring connection to a mythic past. The ability of the champagne négociants to formulate a unique style of marketing that would attract a select group of elite buyers as clients and, at the same time, mold potential new clients out of the growing middle classes proved a healthy adaptation to the market conditions and consumer culture of the period.

Advertising and promoting the product category of "sparkling wine," which began under the Second Empire in France, became more specifically national with the Third Republic. Desire was created within a nascent public in France and abroad who associated the wine with specific French producers, mainly through brands, and an ambiguous yet understood region, mainly through use of the appellation *champagne*. By caricaturing traditional symbols, like those used in champagne ads, as Aaron Segal has demonstrated, "advertising transferred authority and legitimacy to brand name goods defined as products characteristic of France."[131] Efforts to advertise and promote champagne spread common symbols and turned the particular into the national. What were once regional specialties became national goods. It was the success of these efforts that assured that champagne was a crucial part of a transatlantic consumer culture, and, in the words of Raymond Poincaré, a part of "the [French] national fortune."[132]

CHAPTER Three

INDUSTRY MEETS *TERROIR*

Champagne Producers
in the Marne, 1789–1890

"All is lost, save honor," declared René Lamarre in his widely circulated pamphlet *La Révolution champenoise*. Lamarre, a nineteen-year-old champagne vigneron from the town of Damery, echoed the famous words of one of the nation's great kings, written at one of France's darkest hours. François I defiantly penned similar words in a letter to his mother, Louise of Savoy, after his humiliating defeat by Hapsburg troops at Pavia in 1525. He was about to begin his year-long captivity in Spain. For the great king François I, all was lost "save honor and my life." The position of the vignerons producing one of France's greatest wines was no less desperate in 1890. All had been lost, the ardent Lamarre wrote, "save honor and the reputation in this great name—*Champagne*." Much like the foreign threats that François I had faced in 1525, the current danger for Champagne was from an outside force. Lamarre continued, "It is this name, *Champagne*, at this very moment, that they attempt to wrest from you. I cannot repeat it enough: with the way that [wine] lists are drawn up today, *within ten years, people will no longer be acquainted with the name Champagne*, but with those of Roederer, Planckaert, Bollinger, and it will not matter from which [grapes] these wines are produced." Released on the eve of the harvest of 1890, Lamarre's pamphlet was a call to arms for the almost 18,000 vignerons of the Champagne region. The enemy was among them.

The enemy promoted brand names; the enemy distanced the wine from the soil. The daily labor of the vignerons on the land, according to Lamarre, was the source of champagne's reputation. Brand names usurped that reputation and a French national treasure along with it. To halt the assault, the young vigneron proposed a vast project for the "monopolizing of the name Champagne." A united producers' cooperative would allow vignerons to market sparkling wine directly to consumers. Instead of selling at 8 or 10 francs per bottle, champagne would sell at "25 francs per bottle, of which 23 francs will go to the vignerons." The vignerons would turn the regional economy into "the most flourishing in all the land."[1] Unable to buy grapes from the peasant producers, the enemy—with no direct connection to the terroir—would be forced out of the sparkling wine market. Peasant grape growers would save Champagne by linking its famous sparkling wines, not to brand names or a production process controlled by négociants, but, first and foremost, to their place of origin, where the grapes were grown. Champagne would be reunited with its terroir, the true source of its fame. "You are the possessors of an unparalleled product and a universal reputation," he argued, paraphrasing the abbé Sièyes's famous revolutionary pamphlet *What Is the Third Estate?* "And until now, what have you been?—Nothing."[2] Terroir made the vigneron everything.

Terroir is a much-debated term within the wine industry even today. Terroir has been generally applied as describing "the whole ecology of the vineyard: every aspect of its surroundings from bedrock to late frosts to autumn mists, not excluding the way the vineyard is tended, nor even the soul of the vigneron."[3] This combination of soil, topography, and climate are said to create a unique terroir that, according to the *Oxford Companion to Wine*, "is reflected in its wines more or less consistently from year to year, to some degree regardless of variations in methods of viticulture and wine making."[4] Each plot of vines and, ultimately, each region has a unique terroir that creates distinct wine characteristics. The precise conditions of each terroir cannot be duplicated and, by extension, neither can the wines that each produces. The contemporary science of terroir, with the study of microclimates and soils, while still the subject of debate, supports claims, like those of Lamarre, regarding the primacy of the natural environment in determining wine quality. It is, however, the almost mystical quality—"the soul of the vigneron"—in the definition that has made the subject controversial and still creates debate among viticulturalists, historians, and geographers.

Terroir was, first and foremost, used in French to express a geographically imprecise but nonetheless understood sense of region (*pays*) or territory (*espace de terre*) with no specific reference to wine. The main emphasis was on the soil of a more or less understood yet vague area. The more specialized use of terroir for wine dates to the end of the thirteenth century, when it was employed to designate the aptitudes of various soils for the production of grapes. Main emphasis was placed on the role of the soil, particularly its role in governing water supplies to vines, and its interactions with other elements in the natural environment, such as sun exposure. It is through this relationship to wine and more precisely the linking of soil with the subtle tastes of wine that the word *terroir* appears to have entered into popular usage. Terroir became part of nuanced sensory expressions, such as *goût de terroir*, or flavor of the terroir, after the sixteenth century. Wine style and quality were said to come from terroir. By the time that the Académie française took up a discussion of the word in 1740, it had gone beyond linking soil and wine and was applied metaphorically to describe a host of qualities or defects not only in wine but also in people originating in certain *pays*. This transformation was in keeping with the evolution of French medical practice, which placed great emphasis on the physiological effects of wines on character and "personality" disorders such as melancholy or malaise. Indeed, on the eve of the French Revolution, the Royal Society of Medicine in Paris had organized an intensive network of observers in the provinces to make recordings, three times a day, of the environmental conditions of the terroir and the appearance of disease or changes in public health.[5]

By 1800, terroir had become a crucial element in identifying, articulating, and promoting the connection between France's collective health and genius and her material world. Popular assumptions about terroir-related personality traits and prevailing medical practice merged with the new science of gastronomy in this era. An intense interest in the science of gastronomy developed by the 1820s. Practiced by philosophers, encyclopedists, and technicians, gastronomy put even greater emphasis on the role of food and drink in transfusing the human soul with the pure essence found in terroir. Taste professionals in France, like Brillat-Savarin and Grimod de La Reynière, enumerated the qualities of wines and other products of the terroir in books, pamphlets, and journals. One of the fundamental truths of gastronomy, according to Brillat-Savarin, was that "the fate of nations depends on how they are fed."⁶ Romantic-era gastronomy linked the combined contribution of food and wine not only to a diner's mental and physical health but to the health of the nation as a whole.

This focus on national health and the products of the unique French terroir took on a particular urgency with the festering sense of shame over the loss of French territory to Prussia in the 1870s. The fate of the nation appeared to depend on drink and its infusion of the essence of terroir. Wine became the *boisson hygiénique*, the key to both personal and national health. Scientific and medical studies concluded that the best solution for rejuvenating the nation and restoring French values was to create stricter state controls of distilled beverages, like spirits and absinthe, while *increasing* distribution of wine.⁷ A regularly cited study of the French military garrison in Bordeaux, for example, concluded that soldiers who marched after drinking wine were "less tired and went along the road singing and chanting refrains in cadence." This was in sharp contrast to beer-drinking soldiers, who were "sluggish, marched with a heavy step . . . and reached the finishing point worn out, exhausted."⁸ In light of this scientific evidence, the Academy of Medicine endorsed the increase in wine consumption. The medicalization of society had given the early "science" of gastronomy new authority. Far from being a cause of decline and social disorder, French wine and its terroir were deemed the source of national renewal.

As the nation was increasingly urban and industrial over the course of the nineteenth century, the emphasis on the rural, provincial roots of terroir became more pronounced. Writers, such as Balzac in *Les Paysans* (1830), sought to delineate the "natural" state of France, with its symbiotic relationship between people and land. By the end of the century, these ideas were further reinforced by practitioners of the science of geography in France, as articulated by its most revered practitioner, Paul Vidal de la Blache. Central to Vidal de la Blache's classic works, most notably his *Tableau de la géographie de France* (1903), is the view that environment, shaped by internal and external factors, determines the way of life (*genre de vie*) of a locality and its people. According to this logic, a country can only be understood through its environment as determined by its geomorphology. France's geography, according to Vidal de la Blache, situated it at the crossroads of Europe, the crossroads of "the civilized peoples." France was more than the historical synthesis of all these civilizations, it was the culmination of civilization. Vidal's formulation of the authentic, organic France and its terroir, one historian has noted, "clearly owed something to nongeographical considerations."[9] Terroir became, in many ways, the repository for the accumulated historical memory of France; wine became the realization of a resulting French *esprit* in the material world.

While the belief in the organic, animating spirit of terroir became more widespread in the nineteenth century, the connection between the foods and wines available in the capitalist marketplace and terroir became more tenuous. The historic link between terroir and food and drink can still be seen in the names of many French food products. Names of French villages, provinces, or *pays* were frequently used to designate agricultural products considered local specialties, linking them to terroir. By the end of the nineteenth century, improvements in transport, new processing techniques, and reduced production costs made it possible to market these formerly local specialties to an ever-broader group of consumers at home and abroad. Some of these products—roquefort, champagne, cognac, brie—became well known in areas far removed from where they were produced. Culinary experts lauded these foods as vital to the art of eating and drinking. Menus at restaurants, from the Grand Hotel in Monte Carlo to the Ritz in London and the Petit Moulin in Paris, prominently featured regional agricultural specialties.[10] Previously local specialties had become a part of the culture of ingestion by the turn of the century.

The growth of capitalist food production, with the use of industrial technologies and the addition of chemicals and other "falsifications," as they were called, raised fundamental questions about what was, in the end, an agricultural product. New marketing and distribution methods further distanced the consumer from the production process and, more important for producers, like those growing grapes for France's most prestigious wines, the place of origin—the terroir. As these products moved into the mass consumer culture of the period, they risked losing their association with their local place of origin. The name of the village or province could become, in the eyes of consumers, little more than a generic label specifying a type of food or drink. No longer rooted in the local, the name could be used to denote a product category. This can seen most dramatically in the champagne industry. By the Belle Époque, the name "champagne" was used to describe everything from sparkling wines of the Crimea to beer produced in Milwaukee. Indeed, the Syndicat du commerce of Champagne took action against the Miller Brewing Company to stop its advertisements that referred to Miller as the "champagne" of bottled beers.[11]

While the product might be seen by consumers, particularly abroad, as "French," the more specific connection with the land and the people who devoted their lives to working it was increasingly abstract. This was notably true in Champagne, where vignerons had long ceased to produce wine themselves, and the final, intricately blended product had no clear connection to an estate or vineyard, as was the case in most other fine wine regions. Brand names further complicated the matter by promoting "authenticity" based on the manufactured product, not on the origins or purity of the basic ingredients. René Lamarre understood this. It was the grapes, according to Lamarre, carefully cultivated by generations of peasants in the climate of the Marne, not the négociants based in the towns, that made champagne a high-quality beverage. The soil, sun, soul combination at the heart of terroir, rather than the process and the brand names so coveted by the consumers, assured the wine's reputation. By monopolizing production and promoting brand names through the Syndicat du commerce, Lamarre argued, the négociants severed the wine from the heart and soul of Champagne, unjustly accruing all of the regional riches in the process. Terroir was being sacrificed to greed.

Not surprisingly, the négociants, who so carefully marketed their product and aggressively promoted their brand names, scoffed at young Lamarre's cooperative project. His left-wing rhetoric notwithstanding, Lamarre's plan to raise prices to gouge consumers seemed to reveal the naked self-interest behind the project. His accusations about the lack of authenticity in the champagne industry, however, struck a nerve. Négociants struggling to retain control of the appellation "champagne" as industrial property through the Syndicat du commerce counted on their ability to claim an exclusive link to an organic region of France. Lamarre, claiming to represent the "soul" of the terroir, jeopardized these efforts. The Syndicat generated publicity that stressed the role of the wine houses in preserving the authentic "champagne." Age-old manufacturing techniques transmitted from Dom Pérignon, the "father" of sparkling wine, transformed high-quality, regional grapes into the bubbly libation. When the grape growers themselves were mentioned at all, emphasis was placed on the mutually beneficial relationships between négociants, vignerons, and salaried vine and cellar workers. Those who worked for the négociants were praised: "The morality and temperance of those men, their love for their work, and devotion to their employers are widely recognized."[12] The vignerons, it was argued, shared with the négociants in the general prosperity of the champagne industry.[13] While the négociants were practiced at taking the offensive when faced with challenges to their control of the "true" champagne, they were not accustomed to finding themselves on the defensive, facing off with a challenger from inside the industry, particularly a representative of the vignerons, so frequently portrayed as allies. Lamarre painted a menacing picture of "outsiders" who sought to conquer and destroy the organic Champagne and with it the relationship between soul and soil.

Although Lamarre's proposed cooperative garnered only limited support (perhaps, in no small part, because of its utopian nature and because the négociants allegedly boycotted vignerons who showed an interest in the cooperative project),[14] his ideas and rhetoric captured the growing tensions within the wine industry of the Marne in the 1880s. There had always been inequities between négociants and vignerons, but their social and economic relationship before what became known as the *crise sociale* of the 1880s had seemed mutually beneficial. Both négociants and vignerons prospered with the growth of the domestic and foreign markets for champagne, particularly between 1860 and 1880, the "golden age" of French viticulture. While the prosperity of the "golden age" produced a number of wealthy grande marque négociants, it also created a core of prosperous small-scale peasant producers. Like the négociants, Lamarre and the vignerons shared a vision of community based on the commercial production of sparkling wine. As with the négociants, too, there was a widespread belief that champagne was the unique property of a distinct community within France. But as the publication of *La Révolution champenoise* and growing tensions during the *crise sociale* demonstrated, vignerons and négociants sometimes differed fundamentally over whether the "soul" resided with the grape producers or with those who possessed the power to transform the grapes. This was more than an esoteric question; it was a struggle for control of the economic and cultural legacy of champagne.

"The *vignerons champenois*," Lamarre instructed, "possess all the capital: labor and property."[15] The vignerons did indeed own most of the vine land in Champagne throughout the nineteenth century, working their plots with family labor and the occasional hired hand. Scattered across 423 communes within the Marne, 75 percent of these vigneron families had an average holding of less than two hectares.[16] Among the 119 wealthiest communes in 1880, average holdings were even smaller: 7,998 proprietors possessed less than 1 hectare; 2,581 owned between 1 and 5 hectares; 84 had between 5 and 20 hectares; and only 21 owners had more than 20 hectares.[17] This area of vineyards, large and small, was once described by Jules Michelet as an "expanse of white chalk, dirty, poor . . . a sea of stubble stretched across an immense plain of plaster." The underlying seam of chalk that gave the region its plasterlike appearance was fissured and acted like a sieve, draining water away from the thin topsoil and making much of the region ill-suited for agriculture. Within this enormous, flat plain of chalk, there was a small area of steep hills, referred to as the *falaises de champagne*. These hills, with a unique layer of chalk and tertiary debris, were covered by a thin layer of topsoil that was only six inches deep in some places. The thin soil was easily taxed of nutrients or eroded by the cultivation of most crops. Vine planting, however, stemmed erosion and, with the addition of fertilizers, prevented the sterilization of the soil. Possessing even the smallest parcel of this land was for the peasants in Champagne, like peasants throughout France, a family priority. Agriculture specialists in the Marne lamented the tendency of local peasants to cling to their tiny plots. With the average vigneron needing a minimum of 2 hectares to support a family by 1879, agricultural experts reported that "subdivision [*morcellement*] has become excessive."[18]

Historians generally agree that the extreme subdivision of vineyards in Champagne dates back to the eighteenth century. There are no comprehensive surveys of the sale of *biens nationaux* (nationalized properties of the Church, the royal domain, and émigrés auctioned off by the revolutionary regime in the 1790s) in Champagne or the distribution of landholding in the nineteenth century. Because landholding patterns appear to be similar to those found by Robert Laurent for Burgundy, historians have assumed that the Champagne region followed the same pattern following the French Revolution.[19] Most vignerons owned some vines, but not enough to be fully self-employed, while urban elites controlled small areas of vines that comprised some of the best regional vineyards. Vignerons supplemented their income working for other vine owners on a seasonal or yearly basis.[20] Strikes were frequent, and relations between vine dressers and employers have been characterized as "venomous" on the eve of the Revolution. The local *cahiers de doléances* are filled with exaggerated claims of privileged outsiders buying up large tracts of the best vineyards.[21] Once the Revolution began, the vignerons were eager to see the lands of nobles confiscated and Catholic monasteries dismantled. Indeed, the famous monastery of Dom Pérignon in Hautvillers, where peasants were described as "more wretched and poor than anywhere else," fell victim to revolutionary fervor.²² The parceling out of the vineyards continued with the sale of biens nationaux. When divided up into small parcels, tracts of regional vines were affordable by vignerons. In the town of Damery, for example, the biens nationaux were divided into 107 plots, the largest of which was one hectare and the smallest fifty-three *ares*. Most peasant buyers purchased land after it was subdivided into plots of less than ten *ares* each. The properties of émigrés (*biens nationaux de seconde origine*) were also confiscated and sold. These, unlike monastic properties, tended to be scattered among multiple communes. The properties of the Dargent family scattered around Aÿ, Ludes, and Verzenay, for example, were divided among nearly 128 buyers.²³

Evidence suggests that much of the best-suited land, however, was too expensive for vignerons, which is consistent with the evidence of sales in other fine wine areas. These vineyards went to a few urban elites—the ones that the vignerons had so bitterly denounced in the cahiers de doléances. Vine growing and wine production this far north in less than ideal conditions proved too risky for all but a few experienced bourgeois buyers. Buyers willing to assume the risks looked for parcels with the highest quality vines, whose wines had reputations that could garner prices seven times those of more common vineyards. The Moët and Chandon families, for example, bought some of the region's finest vineyards during this period, which became the basis for a sizable wine-producing estate by the end of the nineteenth century.²⁴ In this way, the mixing of elite and peasant proprietors in the Champagne vineyards continued after the revolutionary upheaval. This was counter to the trend in areas of upper Beaujolais, for example, where many members of the urban middle class, rural bourgeoisie, and regional nobility purchased vast vineyards and displaced small peasant proprietors.

The availability of additional land for purchase did not result in an overall increase in the surface planted in vines in the Marne, the area most favorable for vine cultivation. The number of hectares under vine cultivation actually declined slightly by 1840. Only the Aube and Haute Marne saw an increase in vine planting until 1840. Most new vines in these areas, however, were low-quality, high-yielding plants, intended for production of ordinary wines. Indeed, for much of the early nineteenth century, the vineyards of Champagne were known for their still red wines. Most of the 630,000 hectoliters of wine produced each year was not destined to "sparkle" as champagne. Little was understood about the second fermentation process until after 1820, and even then, peasants did not have the resources to produce and market sparkling wines or to withstand the potential financial losses due to breakage of bottles during the second fermentation. Proprietors in some areas—Verzy, Verzenay, and Bouzy, for example—saw their red wines earn excellent reputations among connoisseurs.[25] But these fine wines were only a small part of regional production. Common table wines (*vins ordinaires*) made up five-sevenths of yearly production before mid-century. The proximity to Paris and to regional industrial centers like Reims put local wine producers in an excellent position to take advantage of growing urban markets for inexpensive red wines. The Champagne region "assumed, during the first two-thirds of the nineteenth century . . . a good part of the functions that progressively would be monopolized by Languedoc beginning with the Second Empire."[26]

Even among the champagne-producing vineyards, peasant proprietors continued to need wage employment to supplement their earnings from their small plots. According to Leo Loubère, this became a common practice in vineyards throughout France in the nineteenth century "that complicated the social structure of the wine population and strengthened the symbiosis of large and small growers." There was a general amelioration in relations between the two groups and a "certain sense of comradeship between knowledgeable vine-owners and their skilled workers" that blurred occupational categories within the vineyards.[27] Reporting on vine cultivation in Marne, a proprietor from Oger attested to a sense of respect and comradeship between owner and worker, saying that a good salaried "master vigneron" offered the "protection of a friend."[28] Wages for salaried day laborers increased throughout the early nineteenth century. The rural exodus of the mid nineteenth century, as hundreds of young men and women left the land to work in textile mills, bonnet factories, and other new industries in the area, produced an acute labor shortage in viticulture that resulted in even higher wages for those who remained behind.[29]

Increased wages did not offset the high costs of vine cultivation in the north. The soil was thin and "only produces thanks to the constant care provided and to the furnishing of earth and fertilizer."[30] This meant a continual *bêcherie*, a labor-intensive collecting and ameliorating of eroded soil from the bottom of the gently sloping hills and transporting it back to the top to "re-bed" the vines. Adding to the costs of production were the constant risks of the unpredictable northern climate: frequent and severe frosts in the spring killed flowers and destroyed the chance of a grape harvest; high annual rainfall created humid conditions that encouraged a variety of pests and diseases that left the vines weak during the sometimes dry summer months; cold weather brought hailstorms or caused incomplete ripening of grapes in the autumn months.

Disease was a constant problem for vines. While discovery of chemical pesticides and fungicides helped stem pest and disease infestation after 1850, these applications added to the overall cost of grape production. Oidium, a type of white powdery mildew that grows on the surface of grapes, leaves, and stems, was a major problem for growers throughout France by the 1850s and 1860s. Chemical treatments, although effective, were expensive and labor-intensive. Lime sulfur had to be applied to vines ten to fifteen times a year, using sprayers attached to the backs of vineyard workers. It was made by boiling lime and sulfur, creating a clear, yellow liquid, which was then sprayed onto the vines. This was impractical for large areas, but experimentation in the winegrowing Swiss canton of Grisons led to the creation of a fine sulfur powder that could effectively treat vines on a large scale. It was this early experiment with chemical treatment that led to a "science of sulphur" and a new interest in chemical pesticides.

The cost of such chemicals placed enormous pressure on the household economy of regional peasants. In the early 1850s, "most wine-producers were caught in a scissors-grip between low income and high costs, a situation becoming increasingly common in agriculture as inflationary trends continued."[31] Fifty years later, the cost of cultivation was still high; those who remained on the land could expect annual production costs to surpass a startling 4,000 francs per hectare.[32]

Given the costly efforts to maintain the vines, it is perhaps not surprising that overall wine production in the region declined after mid-century. The decline marked the beginning of a period of transition for local producers, with the gradual phasing out of grape production for still wines and a new dynamism and prosperity for grapes destined for the sparkling wine industry. By 1845, observers noted that the shift away from still wines was almost complete. Although there remained proprietors who were producing still wines, those involved in the marketing of regional wines were exclusively interested in the sparkling wines. According to one merchant, consumers were more inclined to purchase Bordeaux wines

FIG. 1. *L'Accord fraternel*. Bibliothèque municipale d'Épernay.

FIG. 2. Walter Crane, Belle Époque poster, used in railway advertising. Musée de la publicité, Paris.

FIG. 3. "Amour Champagne" (1882).
Archives départementales de la Marne.

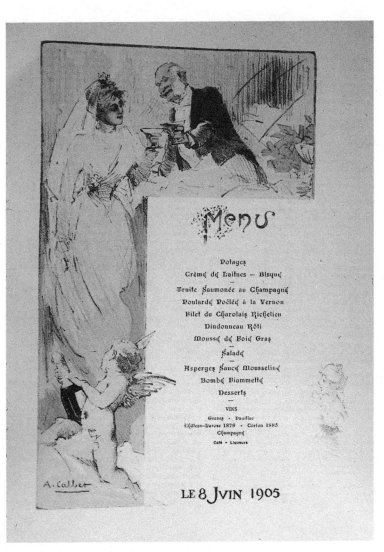

FIG. 4. Wedding menu (1905).

FIG. 5. Wedding champagne (1904).
Archives départementales de la Marne.

FIG. 6. Baby champagne (1886). Archives départementales de la Marne.

FIG. 7. "Toasting the New Year" (1909).
La Vie parisienne, January 2, 1909.

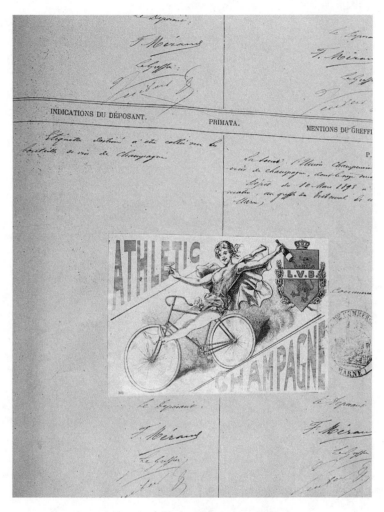

FIG. 8. Athletic champagne (1898).
Archives départementales de la Marne.

Fig. 9. Republican champagne (1876).
Archives départementales de la Marne.

FIG. 10. "Champagne Antijuif" (1898).
Archives départementales de la Marne.

FIG. 11. *Punch* cartoon (ca. 1890s):

"Now, George, my boy, there's a glass of Champagne for you. Don't get such stuff at school, eh?"

"H'm! Awfully sweet. Very good sort for ladies. But I've arrived at a time of life when I confess I like my wine dry."

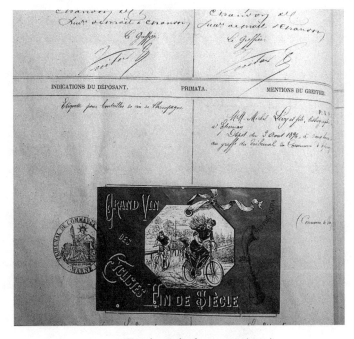

FIG. 12. Fin-de-siècle champagne (1900).
Archives départementales de la Marne.

FIG. 13. Champagne celebrating the 1899 Paris Exposition universelle (1898).
Archives départementales de la Marne.

FIG. 14. Electric champagne (1888).
Archives départementales de la Marne.

FIG. 15. Steeplechase champagne (1889).
Archives départementales de la Marne.

FIG. 16. *The So-called Phylloxera*. Archives départementales de la Marne

FIG. 17. Champagne cooperative poster. Musée de la publicité, Paris.

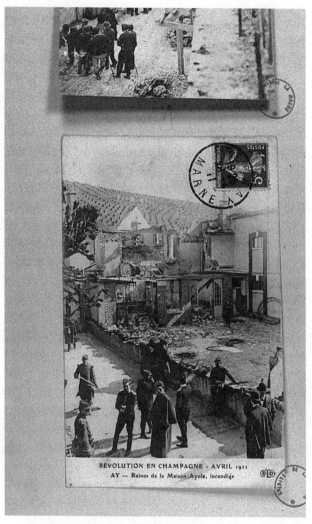

FIG. 18. Revolution in the Champagne region in 1911: The destruction of wine houses. Archives départementales de la Marne.

FIG. 19. Troops in the Champagne region in 1911.
Archives départementales de la Marne.

than regional still wines.³³ After mid-century, sales of sparkling wines soared. Domestic consumption remained relatively unchanged until after 1900, leading one observer to lament that champagne, "the wine of choice, is today in France a royal rarity, like the black pearl or the white blackbird; foreigners pay for it and import it; the French, well placed to obtain the wine, are not certain to drink one bottle of champagne per year."³⁴ Indeed, foreign consumption was the source of the increase in sales, leaping from 3.8 million bottles sold in 1850–51 to 13.8 million bottles in 1870–71. Champagne manufacturers needed quality grapes to meet foreign demand.

For those whose grapes were not destined for sparkling wine, life was increasingly bleak as the century progressed. Local officials at mid-century reported to Napoleonic ministers in Paris that "each day the sons of vignerons are abandoning this difficult work." Evidence suggests that this was particularly true in marginal vineyards along the periphery of the main viticultural areas. "Vines disappear day by day," reported one official from the outlying village of Sainte-Ménéhould in 1858, "because they produce only losses and the finished product has no commercial value."³⁵ Within fifty years, Sainte-Ménéhould had little more than twenty-two hectares of vines producing grapes destined for local still wine consumption. These vines were high-yield varietals that had no "commercial value" because they produced wines of mediocre or low quality.

Champagne wines—whether sparkling or still—were made of juice from Burgundy grapes—the pinot noir, pinot meunier, and pinot blanc chardonnay. The yield of these vines depended on the choice of vine stock, the soil, and the availability and use of fertilizer. Average yield per hectare in the early nineteenth century was about ten hectoliters of wine in areas of quality wine production and twenty hectoliters of wine in areas producing vins ordinaires. Yields were moderately increased in the early part of the century through "amelioration due to replanting with fine vines on hillsides favored by nature" or "imperious necessity favoring, over time, the prolific, more productive vines."³⁶ Average yields for vineyards producing grapes for vins ordinaires ranged from thirty to thirty-three hectoliters per hectare.³⁷ After mid-century, when quality and demand increasingly became the chief determinants of wine pricing, these high-yielding, low-quality varietals were problematic. Reporting on vines planted near Châlons-sur-Marne (today Châlons-en-Champagne), to the east of the main winegrowing areas along the Marne River, for example, an official explained that vignerons were uprooting high-yield varietals because they produced mediocre wines that did not sell. Those who kept these vines, he continued, interspersed potato plants, doing little to improve the overall quality of the wines. The poor quality of wines like those from Châlons-sur-Marne earned them the labels *vins de boisson, vins de pays*, and, less politely, *vins de cochon*.

After 1855, it became increasingly difficult for regional vins ordinaires to remain competitive with similar products from other vineyards. With railroads now connecting Paris and other urban centers with wine-producing regions throughout France, a flood of cheap and plentiful wines, particularly from lower Languedoc, brought new competition. Midi wines were always more abundant and cheaper to produce than those of Champagne and the regions surrounding Paris. The high cost of transport before mid-century, however, added to the costs of Midi wines, making wines from the north more affordable to the working class. Champagne vignerons produced 34.7 percent of the volume of wines of Languedoc in 1840; by 1862, Champagne producers were making only 18.4 percent of the volume of wine produced by their southern counterparts.[38] The sheer volume of Midi wines was not the only problem. Even with the reduced transportation costs after the extension of the Paris-Metz and Paris-Strasbourg railroad lines, which ran through the valley of the Marne, regional still wines could not compete with the lower-priced wines of the south. Costs of wine production in the Marne during the late nineteenth century were estimated at between 1,500 to 2,000 francs per hectare, while in the Midi, costs ranged from 800 to 900 francs per hectare.[39] These costs could not be passed on to the consumer while still keeping the price competitive with that of ordinary red wines from the Midi. Red wine from Hérault sold for 8.82 francs per hectoliter, whereas red wine of similar quality from the Aube sold for 14.86 francs per hectoliter.

High yields of low-quality grapes, coupled with high retail prices and falling demand, created widespread economic hardship in areas devoted to production of vins ordinaires. Vignerons in these areas could convert their land to other crops—cereals or potatoes—or migrate to urban areas in search of work.[40] Many chose to migrate, finding "less difficult work, a higher salary, pleasures."[41] Paris was a particular draw for regional migrants.[42] Out-migration and a declining birthrate among the rural population created a shortage of labor, particularly in the agricultural sector, which plagued the region well into the twentieth century.

Michel Hau further attributes population decline to the general aging of the population and declining fertility rates. Immigration, particularly after mid-century, deprived the region of many men and women of childbearing age. Between 1851 and 1856, the population, which had regularly increased during the beginning of the century, declined from 1,238,200 to 1,212,400.[43] The regional birthrate, which had slowly declined since 1800, was by mid-century lower than the national average. This low birthrate coincided with an increase in mortality, owing in large part to the general agricultural crisis after 1862 and the persistence of cholera epidemics in the region. The population decline and exodus was not as dramatic in the Marne as in the surrounding departments, but the rural population of the Marne fell from 270,400 inhabitants in 1851 to 263,000 in 1856. The slow population decline continued in the early years of the Third Republic. A sharp drop in fertility rates after 1886 kept the birthrate of the region well below the national average. But, above all, it was the rural exodus that was responsible for the demographic decline in the later part of the nineteenth century. The rural population of the Marne, near 270,000 in 1851, had fallen to 215,000 by 1914. Only the towns of Reims and Troyes saw a slight population growth during these years.

Labor shortages, in turn, resulted in high wages for the salaried vine tenders who remained. The average wage of an agricultural day laborer in 1862 (without food or lodging) was much higher than the average in France as a whole: 2.30 francs before harvest and 3.28 after harvest in the Marne, compared to the national average of 1.85 francs before harvest and 2.77 francs after harvest.[44] High wages alone could not entice the sons and daughters of vignerons to remain on the land. In some areas, it was reported that young women refused to marry vignerons. With the textile centers not far away, these young women preferred "to abandon the work of the vine for that of the needle."[45] By 1858, the labor shortage was considered critical, further elevating salaries and alarming officials. A report commissioned by the prefect of the Marne showed that "the dearth of labor is felt particularly in vine cultivation."[46]

This pattern of economic decline and rural exodus was not particular to Champagne. Historians of rural France have noted these same trends in other regions between 1840 and 1870.[47] After 1850, the growth of urban centers in France meant a growth in markets; prices for agricultural goods rose accordingly. Historians argue that the peasant farmers who remained in these agricultural regions were able to meet growing urban demand through better production techniques, a decrease in demographic pressures, and improvements in transport. Moreover, in wine-producing regions, particularly between 1860 and 1880, there was a general movement toward vine monoculture and economies of scale. Capitalist market relations were transforming the rural landscape.

Champagne experienced a similar reorientation toward urban markets and vine monoculture. Economists estimate that this transformation of rural Champagne was much more rapid and dramatic than in other wine regions, however, involving a radical shift in the orientation of the regional wine industry and grape cultivation. With the almost total collapse of ordinary wine production, the area of Champagne *viticole* centered around three main production areas: the vines along the Marne River between Damery and Bouzy; the vines situated on the slopes of the "montagne de Reims," or mountain of Reims, to the south of that city; and the area of superior white wine–producing vines along the hills of Avize, Cramant, Oger, and Le Mesnil. The medieval economic capital of the region and former center of the wine trade, Troyes, was now peripheral to the centers of industrial wine production, Reims and Épernay. While producers of vins ordinaires were struggling or abandoning their land, those planting or cultivating grapes that could be shipped to these centers for use in regional sparkling wines entered a period of prosperity. The boom in champagne sales caused by the general prosperity after 1855 brought increased demand for grapes to be used in the blends that formed the base of the regional luxury wines. To maintain the cellars of regional champagne firms, which contained nearly 40 million bottles of sparkling wine by 1870,[48] négociants "enlarged the circle of production" by purchasing wines or grapes from vineyards scattered throughout the Marne.

The general area deemed suitable for fine wine production—the terroir that gave the regional wines their distinction—had not changed by the mid 1850s. Little could be done to change the thin layer of chalky soil that covered the famous limestone caverns of the region. And despite years of experimentation with methods of protecting vines from the ravages of the climate, the grape growers remained at the mercy of the forces of nature.[49] The shift to fine wine production and vine monoculture did not lead to an extension of vineyards. Rather, the total area of vines within the Marne declined from 17,000 hectares in 1818 to about 14,000 in 1868.[50] Those who survived the transition to vine monoculture found that, despite the difficulties involved in vine cultivation in the northern vineyards of Champagne, "the production of quality grapes commanding a high price made commercial viticulture worthwhile."[51]

The general conditions of prosperity hastened the rise of the prosperous peasants, the peasant "aristocracy," as one historian has called them, who cultivated land in the valley of the Marne, situated between the towns of Damery and Bouzy, the mountain of Reims, running northward toward the city of Reims, and the white grape region on the hills between the towns of Chouilly and Vertus. For these growers, rising costs were easily absorbed by the rising prices for grapes. As one observer noted, few vignerons in these areas lamented the demise of still wine sales, because purchases by négociants "provide all classes of proprietors of vines a prompt and profitable placement of their harvest."[52] Good weather conditions between 1855 and 1880 produced above-average harvests (except in the years 1860, 1866, 1871, and 1879).[53] Growers, particularly in the valley of the Marne and mountain of Reims areas, augmented their production by planting additional vines adjacent to their existing vines. To achieve this, vignerons planted *en foule*, burying branches of a mature vine in a circular pattern around it. These would eventually root and create a separate vine adjacent to the original one. The propagation method led not only to a high density of vines but to nonlinear, "disordered" vineyards that could not easily accommodate animals or machinery. One observer noted that, despite dependence on manual labor and the menace of vine diseases, vignerons in Champagne *viticole* were "transforming arable or uncultivated land into vineyards."[54] In the commune of Sillery, on the edge of the mountain of Reims, for example, only 50 hectares of land were devoted to grape growing in 1830, but by 1893, taking advantage of the increased demand for grapes, vignerons cultivated over 110 hectares of vines there, despite considerable losses to phylloxera.

As with most wine regions, the price of land varied according to the prestige of the wine produced. In the most reputable areas—Aÿ, Sillery, Mareuil, Épernay—prices were four to five times higher than those paid for land in other grape-producing areas. By 1881, it was estimated that the average price of vine land in the arrondissement of Épernay was 7,200 francs per hectare, and in the arrondissement of Reims, the price per hectare was an average of 10,000 francs.[55] One peasant in Ambonnay, "absorbed in the refreshing pursuit of turning over a big heap of rich manure with a fork," informed an interviewer that "every year vineyards came into the market and found ready purchasers at from fifteen to twenty thousand francs the hectare."[56] Vineyards in neighboring Aÿ sold for an average of 25,000 francs per hectare, and, in the best wine regions in Bouzy, land sold for an average of 30,000 francs per hectare.[57]

With land prices climbing, new land purchases remained out of reach for the average family, and holdings among vignerons remained small. A survey of landholding in Aÿ, located almost in the center of the fine wine–producing area between Épernay and Reims, shows that the majority of peasants (333 of the 417 landowners) had less than 2 hectares of land in the 1890s. Most of the "big" landowners owned between 3 and 8 hectares, but two, Raoul Chandon de Briaille and Auban Moët of Moët & Chandon, had estates of 44 hectares and 58 hectares respectively.[58] Overall, in the department of Marne, 17,739 proprietors possessed 15,538 hectares of vines by the mid 1890s. It was estimated that 14,430 of these proprietors had less than one hectare.[59] The vast majority of these landholders, moreover, even among the largest estates, lived within the region, most in villages of 500 to 1,000 inhabitants.[60] This was a remarkable density of population compared with non-grape-producing villages in the same area, which had an average population of 200 inhabitants. Semi-urban villages coupled with extreme subdivision of land (particularly compared to other areas of fine vine cultivation, such as Burgundy, where the average vine-tending household cultivated two or three hectares) gave the region a certain peculiarity.[61]

Small growers in these areas could hire themselves to large landowners as day laborers or *tâcherons* (pieceworkers for a specific task), commanding high wages that supplemented family income.[62] Labor provided by these small growers was increasingly in demand after mid-century. A number of historians have pointed to this period as one of accelerated growth of large-scale capitalist agriculture in the region. They argue that it was the négociants and their families who bought land and consolidated their holdings in fine wine–producing areas, creating a very productive industrial-style enterprise. With the phylloxera epidemic in the 1880s and 1890s, in particular, "the ruin of numerous small proprietors was the occasion for the négociants to create vast vineyards and to initiate a movement toward vertical integration."[63] Because purchases of property traversed hundreds of small communes, it is difficult to get an accurate picture of the rate of increase in landownership and purchasing patterns among négociants. Accounts by contemporaries suggest that négociants were indeed buying land and establishing their own wine presses in viticultural communes. Their ability to pay high prices and the size of their holdings set them apart from the majority of the proprietors. Moët & Chandon, for example, had some 365 hectares of vineyards by 1900, distinguishing the firm from even the largest peasant landowners in a region where the average peasant cultivated a single hectare.⁶⁴ Moreover, the scale of these négociant-controlled vineyards allowed them to experiment with production techniques, including the gradual adaptation of linear vine planting, that facilitated cultivation, increased yields and made application of pesticides and fertilizers much less labor-intensive.

The growth of this large-scale capitalist viticulture did not destroy small producers. On the contrary, the growth of large estates created an important source of supplementary family income for the peasant producers. From planting to harvest, the yearly cycle of vine tending in Champagne was above all labor-intensive. Large proprietors were unable to complete the cycle without skilled laborers, who often came from the ranks of small proprietors. During the early part of the century, proprietors hired tâcherons who worked at specific jobs in the yearly cycle of labor for a fixed price. After mid-century, however, those seeking wage labor generally met at a central *bourse de travail*, or labor exchange, where proprietors hired day workers (*journaliers*). Large landowners, like Moët & Chandon, found it necessary to hire over 800 vine tenders per year by the 1890s, not including those needed to complete the harvest.⁶⁵ Many who requested *livrets d'ouvriers* to work in the vineyards or wine houses in the early 1890s simply listed *chez plusieurs* (with several) as their place of employment, indicating the fluid nature of employment in the region, but by the early 1900s, the place of employment was increasingly a single maison or négociant-owned vineyard.⁶⁶

Evidence suggests that wage labor allowed many vignerons to weather years of poor harvests, incorporate costly treatment programs, and continue to cultivate their smallholdings.⁶⁷ Because agricultural laborers were often hired on a daily basis, it would be difficult to estimate how many vignerons' families were combining wage labor with landownership by the end of the nineteenth century. Archival records indicate, however, that increased reliance on wage labor created new occupational categories among the rural population. Whereas there had been little distinction between salaried or unsalaried vignerons earlier in the century, after the 1880s, officials, négociants, and journalists, as well as peasants and other members of the rural population, made distinctions between peasants who primarily worked their own land (*propriétaires-vignerons*) and those predominately engaged in wage labor (*ouvriers-vignerons*). Large landowners, by contrast, were distinguished by the title of *propriétaire*, or, if they were from among the champagne-manufacturing families, *négociant*. Although peasants often listed their occupational category as *vigneron*, the term had come to denote a "simple peasant who trimmed vines regardless of whether they belonged to him or to somebody else."[68]

The title of *vigneron* also denoted status and skill. Male vignerons were considered the most skilled and earned on average around 5 francs a day (or 1,000 francs a year, based on 200 days of employment), putting them among the highest-paid wage laborers.[69] But there were also skilled female vineyard workers, *vigneronnes accomplies*, whose status in the community was frequently recognized by the hyphenation of their family names with those of their husbands, and who often took over family vineyards when widowed. Yet they were not judged by employers to have sufficient discipline for highly paid skilled labor. On average, they could earn 600 francs a year, based on 200 or 220 days of employment.[70] Children performed a number of unskilled tasks, working in the vineyards for eight to ten hours daily. Estimates put the earnings of children over the age of fifteen, working 200 days a year (including the harvest), at 500 francs per year.[71]

The general prosperity in Champagne between 1855 and 1880 led a number of commentators to report that the vignerons of Champagne were "all rich and comfortably well-off."[72] They had a lucrative market selling grapes directly to the négociants, although prices could vary considerably from vineyard to vineyard and from year to year.[73] In the late 1870s, there were peasants who sold the "bulk of the vintage . . . before a single grape passed through the winepress," Henry Vizetelly reported. "One peculiarity of the Champagne district is that, contrary to the prevailing practice in other wine regions of France, where the owner of even a single acre of vines will crush his own grapes himself, only a limited number of vine proprietors press their own grapes."[74] Instead, "the multitude of small cultivators invariably sell the produce of their vineyards" to the large wine houses or large landowners. In areas of quality grape production, vignerons could count on representatives of the large wine houses purchasing the bulk of their grapes while still in the vineyards and hauling them away for crushing.

Négociants could own multiple wine presses near their own vineyards or in the areas where they made frequent purchases. In areas where the négociants were most interested in obtaining grapes, they sometimes offered growers lucrative yearly contracts to deliver the harvest directly to the presses. Prices for regional grape production from the 1870s until the early 1880s rose steadily. The Syndicat du commerce's records show that prices for grapes in the communes of Bouzy and Avize, for example, increased from 400 francs *la pièce* (205 liters, or enough grapes to make 273 bottles of champagne) to over 900 francs between 1863 and 1873.[75]

One of the most prominent historians of the champagne industry, François Bonal, has argued that the négociants imposed the practice of direct sales on the vignerons. Economically, he explained, it was easier for the négociants to obtain an advantageous price for a perishable item, like grapes, than for a product that was already finished, like wine, which could be stocked for future speculation. Moreover, by purchasing the grapes directly, the négociants could follow the development of the wines from the moment that the grapes entered the press. Supervising the winemaking from pressing to first fermentation gave the négociants greater control over the quality of the wines long before the blending and bottling for the second fermentation process began.

While the practice of direct sales was advantageous to the négociants, it would be difficult to support the contention that this arrangement was "imposed" on the vignerons. Arguably, the system worked to the advantage of the négociants over the long term. Direct grape sales, however, appeared initially to free vignerons from some of the yearly uncertainty and costs of wine production and sales. As one educated vine grower wrote, the vigneron "waits for December between fear and hope. Will the wine sell well? Will the wine sell badly? Will the wine sell at all? *Hélas!* ... He has harvested well, pressed well, and carefully developed his wine; but, in spite of all that he has done, will this wine please the négociants?"[76] The yearly crop was sold before or at the time of harvest, putting money in the hands of the vigneron family much more quickly than under the previous system. Moreover, according to the journal *La Vigne*, the system of direct sales "has exempted the *petit vigneron* from the costs of pressing, barrels, etc., and each has found it profitable."[77] Direct sales reduced overall production costs for the peasant household. In areas where the "vigneron never made wine himself," families could abandon wine presses and storage sheds and eliminate the labor of winemaking.[78]

Freed from "the concerns of a delicate manipulation," vignerons focused on increasing yearly revenues by maximizing grape production.[79] In this effort, the local vigneron paper noted in 1873, there was a "tendency to replace quality with quantity since the establishment of the system of buying grape harvests."[80] While increasing plantings, either by increasing the density of vines or transforming every available bit of soil into a vineyard, vignerons also lavished attention on existing vines, employing new fertilizers in an attempt to generate higher yields. With average yields at a modest 24.5 hectoliters per hectare and demand for champagne continually rising, every grape was precious.

❦

Harsh weather conditions and adversities such as oidium, mildew, various fungal diseases, and phylloxera constantly threatened the livelihood of the vignerons in the late nineteenth century. The diary of one vigneron chronicled this constant struggle: in 1872, "all the vineyards of Champagne experienced a frost in spring, except Venteuil; 1873, a frost, as in midwinter, which was general, destroyed the harvest by three-quarters (April 25, 26, and 27); 1874, frost destroyed up to three-quarters [of the harvest]; from May 1 to 22, there was a frost more or less each day."[81] While chemicals proved effective against a host of diseases and insects, there was little that could be done about the climate. By the late nineteenth century, there were several types of shelters (*abris*) that were used to cover vines during periods of frost or hail. Covers of straw or wood would be rolled across the top of the vine stakes, creating a layer of protection. These were expensive, labor-intensive solutions, however, that were out of reach of the majority of small proprietors.[82] And there were always the latest inventions, such as the *fusée paragrêle,* or hail rocket, advertised in vigneron papers. Hope for a miracle led to fantastic claims: launched into a hail cloud, the rocket would allegedly dissipate the ice before the deadly projectiles were unleashed.

Organized efforts to deal with these natural adversities, before passage of the 1884 law authorizing professional associations, or *syndicats,* were rare. One *vigneron-propriétaire,* Georges Vimont of Mesnil-sur-Oger, attempted to form a peasant-based organization in his canton in 1878 and 1880 to study the mysterious new disease called phylloxera. The administration declined his request to do so, however, fearing that such an organization might have political motives. It preferred, instead, to create a departmental commission, the Comité d'étude et de vigilance contre le phylloxèra (hereafter the Vigilance Committee), which, after being officially sanctioned in 1879, held meetings several times a year to monitor the progress of phylloxera in northern France, discuss treatment methods, and review research on other vine diseases. Unlike Vimont's proposed peasant-based organization, the Vigilance Committee consisted mainly of agricultural professors, large landowners, and négociants, and it remained aloof from the mass of peasant proprietors.

Only in the 1890s were Vimont and other peasant proprietors successful in creating peasant-based organizations for dealing with the problems of vine cultivation. These organizations were mainly concerned with either obtaining materials at lower prices or collectively sharing the costs of treatment methods. In 1894, for example, the vignerons of Bouzy joined in a "*syndicat* for preserving the vines against spring frosts" by creating *nuages artificiels*.[83] At the first signs of frost, vignerons ignited a number of small stoves that created a warm, smoky "cloud" above the vines. In 1903, the same syndicat undertook collective efforts against an infestation of moths, and a year later, members experimented with new vine stakes.[84]

Even before the creation of these organizations, however, vignerons had made some informal collective efforts to adjust work routines to maximize protection of the vines. During the months of May and June, known as the *travail fou* (work like crazy) or *travail à quatre bras* (work double quick) season, for example, vignerons labored for long hours so that no one would need to enter the vineyards after the vines had flowered. This new custom was honored throughout the Marne, where vignerons practiced a unique vine reproduction method that could result in as many as 45,000 to 50,000 vines per hectare. Given the density of planting, even a single owner traversing the paths to his vines could disturb or destroy young flowers or newly formed fruit on his neighbor's vines. Only after the fruit appeared did vignerons recommence work in the vineyards. This custom was legitimized as a tradition worthy of legal protection. Some communes, like that of Épernay, for example, passed strict laws based on these "ancient customs," forbidding anyone to enter the vineyards at night.

Regardless of whether vines were cultivated by family labor, hired workers, or a combination of the two, the year-long cycle of fertilizing, pruning, combating pests, and struggling against the elements, known as *les roies*, was the central concern of the vignerons. The family members worked side by side, going out at dawn, carrying all of the tools, food, and wine needed for a day of labor. One photo of a vigneron family working their vines in winter shows a woman, dressed in her distinctive hat (known as a *bagnolet*) bending forward, straining to carry her *hotte* (a basket carried on the back) of compost, while a man and a young boy work the snow-covered field, surrounded by hundreds of dormant vines.[85] Work continued year-round, for eight and a half to ten hours per day, and no one was spared from the more onerous tasks.[86] One observer wrote that the working conditions for the women of Champagne (the vigneronnes) "are scarcely easier than those of men . . . they [women] are true athletes in their strength and their courage."[87]

Work was made more difficult by the system of planting en foule. Regional vignerons, like their counterparts in Burgundy and the Jura, continued to plant en foule until after the turn of the century. The term *en foule* derives from the disordered arrangement of the vines, likened to a "crowd" (*foule*) of people, but it eventually came to signify a "large number," because in this type of growing, the density of vines and shoots is much higher than with vines planted in rows. Cultivation en foule involved the rejuvenation of vines by extending a vine branch from a "mother" plant and layering it with soil and manure. Layered shoots would eventually sprout roots, establishing a new, independent plant, which bore fruit within two years. This layering operation was repeated each year until the original vine was five years old. At five years, the vine would be "rested" for several years before the process began again. Although this system created a "great haphazard profusion of vines [that] resembled a disheveled mass covering the hillsides," it constantly rejuvenated the vineyard, giving a single vine a lifespan of almost seventy years.[88]

The high density of vines and shoots created by this method made it difficult to use machinery or animals during the yearly cycle of cultivation. Mechanized plows and harvesters were incompatible with the haphazard arrangement and density of the vines. Even simple nonmechanized plows were rare in the vineyards, and there were few animals to aid workers with heavy tasks. In the 1890s, vines in some areas of the valley of the Marne and the mountain of Reims were replanted in rows with grafted vines, which had a much shorter life than those planted en foule. Despite these changes, however, there was little mechanization until after the 1920s.

Because of the density and constant proliferation of vines, the wooden vine supports used in the region had to be removed and replaced several times per year. Women and children worked during the winter months removing the wooden stakes used to support mature vines. Stakes were carefully inspected, and any showing signs of bug or disease infestation were treated. The following April and May, all the vine stakes had to be replaced. Multiple supports, often made of pine or oak, were necessary for each vine and could number as many as 50,000 per hectare in some vineyards. Peasants used their feet or their chests to force the stakes deep into the soil. "In former times," one source explained, "only the women *fichaient* [forced vine supports into the ground] . . . the woman [*la ficheuse*] inserted the butt [*talon*] of the stake into a leather breastplate [*plastron*] that she wore on her chest and bore down on it with all of her body weight until she forced it into the ground." By the end of the nineteenth century, however, men and women shared "this dreadful labor."[89]

The shallowness of the soil throughout Champagne *viticole* created another form of "dreadful labor" that was peculiar to the region. Unlike other wine regions, Champagne vineyards required intense efforts to stem erosion and constantly renew nutrients in the thin layer of earth. This was undoubtedly the work being carried out by the family in the photo cited earlier. François Bonal notes that this was "the most important work for [regional] viticulture, work that cannot be found in any other famed vineyard in France."[90] By the end of the nineteenth century, vignerons fertilized dormant vines during the winter months, creating small trenches and retaining pools to capture any soil eroded by winter rains. In the spring, they collected this soil, as well as soil that had built up at the bottom of hills, and mixed it with compost and manure. The mixture was loaded into baskets, which were then carried up the narrow paths, strapped on the backs of family members or workers. There were "rest stations" of sorts—a flat wooden platform on the edge of a path—that allowed the vignerons to rest the bottom of the basket for a few minutes before continuing up the steep path. Once the hauler arrived at his plot, others carefully removed the mixture, placing it around the vines and mixing it with the existing soil.

The harvest at the end of this cycle of continuous labor and care was anticipated with a mixture of hope and anxiety. Unlike in other areas of France, there was no *ban de vendanges* in Champagne that prevented harvesting until an officially proclaimed date. But, because it was believed that very mature grapes were best for producing sparkling wine, vignerons often began the harvest late in the fall. Harvesting, which lasted approximately twenty days, required the hiring of additional unskilled workers. Throughout much of the nineteenth century, migrant workers came to Champagne from the departments of Lorraine and Alsace. With the Franco-Prussian War and the loss of Alsace-Lorraine, however, the yearly influx of harvesters was disrupted.[91]

A *louée locale* (a local hiring fair) was held in each commune for those interested in working the harvest. "Tomorrow at three o'clock in the morning," wrote one proprietor, "the church bell will wake us so that we can go to the hiring place [*place de louage*]. We'll find all the available harvesters [*vendangeurs et vendangeuses*] there."[92] Proprietors selected workers and grouped them in work teams called *hordons* or *hordes*. These teams, consisting of grape pickers, porters of small baskets [*petits paniers*], and pairs of porters for the large baskets [*gros paniers*], steadily moved the grapes from the vines to waiting wagons or pack animals.

Grapes were taken to the large buildings within the vineyard that housed the wine press. But before the grapes entered the press, they underwent a second grape selection known as the *épluchage*, or *triage*. This systematic removal of grapes that were damaged, insufficiently ripe, or overripe was unique to the Champagne region. In the past, when vignerons had made their own still red wines, there had been a similar, although less selective, process that took place in the vineyards, usually overseen by the vigneron or his spouse. With the switch to direct grape sales, however, it was an established practice that "the buyer had the right to demand that the grapes be selected [*triés*] and cleaned as if they were to be served *sur la table*."[93] Women were hired to inspect the harvested grapes when they arrived at the négociant-owned presses, removing "green or rotted grapes rapidly with the end of [their] scissors."[94] Discarded grapes were saved for home consumption. According to a pamphlet put out by the Syndicat du commerce, by the late 1880s, this process of épluchage was the norm where grapes were sold directly to négociants.[95]

Except for vignerons with yearly contracts, those interested in selling their grapes had to bargain with négociants or their agents. Generally, a base price was established for grapes harvested in the mountain of Reims area, where some of the best wines had previously been produced. These prices were posted daily on the mayor's door in each commune and formed the basis for negotiations between négociants and vignerons. Vignerons knew that with direct grape sales, "prices vary according to the quality, the quantity of the harvest and the needs of the trade."[96] Between vineyards, prices generally varied according to the level of quality associated with the basic wines that had previously been made there. This system could create enormous variations, which were considered an injustice by the vignerons of the *petits crus*.

Wine experts were clear about the difference in quality between wines produced, for example, in the vineyards of Bouzy and those of neighboring Ludes. With new plantings and direct grape sales, however, the importance of these distinctions was less clear. Négociants purchased grapes throughout the Marne, and some observers maintained that by the 1880s, "the distinction [between areas] has nearly ceased."[97] Yet, despite grumbling from vignerons in the petits crus, there was no move to reclassify vineyards according to grape quality. Instead, a complicated system of creating an average price for the wines of each commune remained in place until 1911.

Although the system of grape pricing remained the same, a report published in the 1870s hinted at the extent to which market relations between vignerons and négociants had changed. The article reported that in 1873, "several producers, in the hope of selling their grapes by the kilo, did not take the precaution of furnishing themselves with barrels . . . ; their expectations being far from those of the buyers, they are forced today to make wine."[98] This statement, tucked among articles concerning the politics of the early Third Republic, contained nothing that would alarm or shock vignerons. The relatively unremarkable article, however, hinted at the growing dependence of the small grower on the prices offered and purchases made by the négociants. With fewer growers keeping the materials necessary for winemaking, grapes, which could not be stored, had to be sold at the time of harvest. The "needs of the trade" were now central to grape pricing and the peasants' livelihoods.

Altered social relations and evolving market practices developed almost imperceptibly into new "traditions." At the same time, old traditions and peasant rituals were slowly abandoned. "Except in winter or during periods of bad weather, one does not dream of practicing the *veillées*," Raoul Chandon de Brialles wrote in 1897.[99] Veillées were informal village gatherings that usually took place in the evenings during winter months. These were important social events, where dancing, singing, and lively conversation created a festive atmosphere. Changes in work rituals contributed to the demise of the veillées. New fertilizers and chemical applications, for example, increased winter labor and left little time for evening festivities. And the demise of the practice of burning wooden vine stakes increased fuel costs, making the heating and lighting for veillées an expensive luxury. With the end of the veillées, social rituals, like courting and marriage arrangements, became private family affairs, conducted, not under the scrutiny of the village, but during the afternoon periods of rest from vine work.[100]

Village rituals tied to the Catholic Church were also abandoned, such as those surrounding Saint Vincent, the patron saint of vignerons, which had been almost completely eroded by 1900. Traditionally, the two-day celebration began on January 21 with a cannon blast and young men carrying a statue of the saint through the village. Offerings of red or white grape seeds were attached, in alternating years, to the right arm of the statue. After a mass the following morning, the congregation paraded through the streets, before returning the statue to its sanctuary in the church. By the 1890s, the religious celebration had been abandoned, and the village banquet and celebration that usually followed the mass were transformed into a day of private family gatherings.[101] Secular aspects of the celebration, however, were resurrected during the protest movement of 1911. And women continued to celebrate the festival of Sainte Agathe, patron saint of vigneronnes, each February, although by the 1890s, the "festival" was little more than a day of rest after an early morning mass.

Despite the transformation or erosion of traditions, the rural village remained central to the life of the vine-cultivating community. By the early years of the Third Republic, these villages, scattered among the vines, were not the backward hamlets found in other agricultural regions. An average vineyard village in Champagne had between 500 and 1,000 inhabitants, with about 100 people per square kilometer. These villages were densely populated compared to villages in neighboring wine regions, which had an average of about 300 inhabitants. Vineyard villages were successful commercial centers, populated by vignerons—salaried and unsalaried—artisans, small shopkeepers, and, in some cases, wage laborers employed outside the vineyards. In villages in the valley of the Marne, in particular, wage workers moved frequently between the village and the city of Épernay in search of employment. Villages were less the face-to-face communities described by some historians of rural France than thriving economic centers with a "semi-urban character."[102]

The predominance of smallholding peasants and the importance of wine production to the village economy assured that vine cultivation and agriculture were central concerns of the population. In the political realm, in particular, issues of importance to viticulture predominated. According to one study of political rhetoric and campaigns of the late nineteenth century, local problems were largely absent from electoral politics, with the exception of questions of specific concern for Champagne *viticole*. These were not, however, esoteric concerns of local markets. Many of the concerns or grievances were traced to issues outside of the local community—import taxes, trade agreements, legislation regulating wine—and demonstrate an awareness of national politics and capitalist markets. While the champagne négociants supported anti-republican causes and right-wing candidates until the early twentieth century, the rural population, according to one interviewer, showed "unshakable republican convictions."[103] After 1879, "republicanism" in the Marne was above all democratic and egalitarian, grouping vignerons, artisans, wage workers, and *petits fonctionnaires* in rural villages.[104]

Republic, *patrie*, progress, and justice were recurring themes in successful electoral campaigns in the Marne between 1871 and 1914. This was perhaps not surprising in a period marked by the emergence of a national culture and the creation of "national attachments," as Caroline Ford has shown in her work on Brittany. For those in rural Champagne, however, embracing the new republican nation meant that this nation had to welcome the region, with all of its distinctiveness, as part of the national territory. This was key to electoral victory among the politically engaged rural population. Successful political candidates had to prove that they were "Champenois." To be Champenois was to have the characteristics of the terroir, which by the 1860s meant that an individual had to be attached to the rural world to possess the animating spirit of the region. This theme was used effectively in rural elections in the Marne in 1888, for example, when Léon Bourgeois, the "favorite son" (although born in Paris), defeated the "foreigner" (*étranger*) General Georges Boulanger (a native of Rennes, in Brittany). Again when the nation was challenged by internal divisions during the Dreyfus Affair, rural voters responded to candidates who claimed to "share" the life of the rural electorate through military service and attachment to the rural world, proving themselves to be "children" of both the region and the patrie. These were, above all, the themes of Radical party candidates, and, after the overwhelming defeat of Boulanger in 1888, rural Champagne remained a stronghold of the Radicals.[105]

Although larger political issues, such as the Dreyfus Affair and the Panama Scandal, appeared to generate little interest in Champagne *viticole*, neither rural villages nor the vignerons who formed the core of the population were isolated from French society or the capitalist economy of the late nineteenth century. The rural population appears to have been highly literate; over 90 percent of the men applying for livrets d'ouvriers in 1890, for example, were able to sign their names,[106] and rural newspapers were commonplace. With the circulation in 1873 of the first issue of the *Vigneron champenois*, a newspaper dedicated to reporting on issues of interest to Champagne *viticole*, vignerons and négociants had a common source of news and information about the sparkling wine industry and regional issues. While other national and regional newspapers were available, the *Vigneron champenois* addressed the particular needs and interests of the champagne industry. Headlines and articles focused on local, national, and foreign developments of interest to the community. Champagne production created substantial ties that bound the interests of agriculture and industry, vigneron and négociant, region and nation.

The world of sparkling wine production had, however, become far removed from the daily travails of the peasantry cultivating the grapes. Champagne wine production, by the 1880s, was much more highly mechanized and industrial than other wine industries. The sparkling beverage known as champagne did not differ fundamentally from other wines; it was an additional manipulation of the wine that produced the "sparkle" of gaseous bubbles in what were normally "still" regional wines. The process of creating a sparkling wine involved inducing a second fermentation of a nonsparkling, or still, wine. In order to stimulate the sugars and yeasts in already fermented wine, a small amount of fresh wine was added and the mixture placed in hermetically sealed bottles. The gases created during this second fermentation were trapped in the bottle, creating the wine's coveted "sparkle."

The wines that went into champagne generally came from the juice of Burgundy grapes: pinot (noir and meunier), also used in making fine port and burgundy, and chardonnay, often used for fine chablis. There are two exceptions to this general rule: a champagne called *blanc de blancs* made entirely from white grapes, and *blanc de noirs*, a white wine made entirely from red grapes. The blending of these red and white grapes gave the finished wines their own distinctive taste. Blends of wines from the three vine-growing areas—the mountain of Reims, the valley of the Marne, and the Côte des blancs—were deemed essential to achieving a balanced wine. While most of the grapes grown were normally destined for still red wine, local producers developed a method of pressing the red pinot grapes early to obtain a slightly tinted or tawny white wine. By the eighteenth century, winemakers had managed to produce a clear white wine from red grapes, which became the base, making up about three-fourths of the blend, for the fine wines of the region.

Lesser-known *crémants* (slightly sparkling wines) existed long before the industrial production of sparkling wine. The bubbles occurred naturally when wine was made late in the year and the winter cold paralyzed the yeasts that normally turned the grape sugars into alcohol. Warm spring weather reactivated the yeasts, and the fermentation began again, producing a carbonic gas that created a slight "sparkle" in ordinarily still wine. These sparkling wines, a result of what is known as the *méthode rurale*, were generally viewed with suspicion before the eighteenth century.

Evidence suggests that wine merchants (and presumably their clients) preferred the region's red wines, which rivaled those of Burgundy, believing that bubbles were a trick by producers to cover the harsh taste of bad wines.

Efforts by wine merchants and clergy to manipulate a "sparkle" in some of the region's finest wines proved a major success. Local winemakers and merchants revolutionized the industry, "first by shifting from a still to a sparkling beverage, and then to changing its color."[107] Improvements in quality gave the wine a stronger market base with Europe's wealthiest classes, and sales of sparkling wine steadily increased. Pleasant-tasting and effervescent, the new sparkling wine burst into court circles. Sparkling wines made from *vin gris*, although not always favorably received by wine experts, became a passion among the rich and elegant, first at Versailles, and then throughout Europe.

By the end of the nineteenth century, creating champagne—from the blending of the still wines to the second fermentation—had become a complex industrial process. Visits to the "caves de Champagne" became a standard part of the tourist itinerary in France.[108] Visitors from France and abroad marveled at the elaborate mechanized "factories," bustling with activity, and the "introductions of a scientific character" that produced some of the most famous sparkling wines.[109] Workers moved through miles of darkness preparing the millions of bottles of champagne sold each year. Wines were tasted and the sugar content measured by a saccharometer before being blended in large vats (in what was called a *coupage*). After the wines had been aged in wood, a professional "blender" would follow the development of each wine to determine its attributes and the proportions that would be needed to create a balanced blended wine. Wines from various vineyards would be blended with previously aged wines to "marry" certain qualities in a *cuvée*. Blending was a complex, capital-intensive process, involving large amounts of equipment and labor. The proportions of wines from vineyards within the region and older wines remained a carefully guarded secret, because the distinction of each finished champagne came from its unique cuvée.

By April of each year, the manufacturers would begin the process known as *tirage en bouteilles*, placing the wine in bottles, where it would undergo the second fermentation process. For the manufacturers, the three weeks of second fermentation was a period of particular anxiety because of the potential loss from breakage. Scientific discoveries regarding yeasts and sugars and technological applications for monitoring and controlling internal gas pressure resulted in a reduction of loss from 25 percent in the late 1850s to an average loss of about 10 percent by 1870. Once placed in bottles, the wine was left to age for two to four years, depending on its quality and final destination. Ordinary champagne was seldom aged for more than two years. At this stage, before undergoing any additional manipulation, the wine was referred to as *brut*.

During this second fermentation, a small amount of sediment formed on the side of the bottle, which eventually had to be removed. In order to do so without disturbing the wine, bottles were placed, neck first (*sur point*), on racks made of two boards united at an acute angle to create an inverted V. Each day, for six or more weeks, a worker would twist the bottle in a succession of slight turns, slowly moving the sediment toward the cork. When the sediment gathered in the neck of the bottle, it was removed in a process called *dégorgement*, which was the task of skilled artisans wearing wire masks and protective gloves. The *dégorgeur* took "the bottles with which he is constantly supplied from baskets so placed that he can grasp each without a change of position, seizes the bottle with its neck downwards, removes its iron fastening with his *crochet* [iron hook], and the cork, driven out by the gas, comes out with a bang, followed by the sediment and a flow of frothy wine."[110] With the sediment removed, the wine was passed down the "assembly line," where another set of skilled laborers added a liqueur (*dosage*) made from a combination of sugar and cognac. These workers were skilled, not in winemaking, but in operating a small machine with a regulated tap, which added the exact amount of liqueur required.

From there, the bottle was corked on another, slightly larger apparatus, which resembled a miniature guillotine. Corks were soaked or steamed and then placed in a clawed pincher and fitted into the mouth of the bottle. Another worker then drove the cork into the bottle using a small mallet. Wire to hold the cork in place was applied by a series of workers, who also gave the bottles several rapid turns to assure that the liqueur was mixed with the wine. The bottled wine was then sent to workrooms where laborers, generally young women, washed, dried, and labeled the bottles, sorting them into cases depending on the wine's final destination.

Moët & Chandon typified the large producers that dominated the industry in the nineteenth century. With their industrial-style production facilities, clustered mainly along the avenue de Champagne in Épernay and the butte Saint-Nicaise in Reims, these firms were far removed from the small-scale peasant world. Multilevel facilities, which might descend four or more stories underground, and their motorized transport systems, steam engines, and vast networks of cranes and pulleys held a particular fascination for visitors. "In one establishment," Charles Tovey recounted, "the cellars are divided into seven vast compartments, which contain seven subterranean passages ... in which are arranged in order two million and a half bottles of Champagne. The whole establishment covers an area of twelve acres and a half. The total length of the vaults is nearly two miles."[111]

Despite the industrial nature of production, the grandes marques firms, like the small-scale producers, used a combination of family and wage labor. This was a pattern consistent with the organizational structures found among many small and medium-sized French businesses at the end of the nineteenth century. Sons, sons-in-law, and, to a lesser extent wives and daughters, like Madame Heuser of the Mumm firm, nephews, and cousins were recruited and apprenticed in family enterprises.[112] Undoubtedly, the level of production and mechanization and the overall size of the firm played an important role in determining the employment of family members. Names such as Desbordes et fils, Dinet-Peuvrel et fils, Koch fils, Pfungst frères, Fréminet et fils, and Jacquart frères attest to the historic importance of family to the overall structure and succession of the firm. Marital alliances and partnerships between families were often reflected in these names. Firms created or expanded through marriage alliances between négociant families were often marked by hyphenated names: Auger-Eysert et Hatton, Dinet-Peuvrel et fils, Vix-Bara, Chanoine-Ecoutin, Miller-Caqué et fils. It was primarily family businesses outside of the urban manufacturing centers that used hyphenated names, a traditional practice in rural areas of the region, where complex familial relations could lead to confusion in distinguishing firms or property lines, particularly at harvest.

The business structures of the grandes marques firms were, however, becoming more complex at the end of the nineteenth century. Increasingly, firms bore a family name followed by "Cie," indicating a more complex ownership structure. Membership lists of the Syndicat du commerce demonstrate that by 1911, nearly every member firm was a "company."[113] Firms remained family owned; the use of "et Cie" indicated that the firm was operated by a series of interlocking families, some of whom were "silent" partners. Two of the most prominent examples of this were the firms of Heidsieck and Veuve Clicquot–Ponsardin. Heidsieck was directed by the Walbaum family, and ownership rested not with descendants of the founder but with members of the Kunkelman and Piper families.[114] Family members of the original Veuve Clicquot were silent partners of the firm of the famous widow, but by 1913, it was the Werlé family, descendants of Veuve Clicquot's former clerk, who controlled the day-to-day operations of the firm.[115]

The infusion of capital from the expanded circle of partners and increased profits allowed the largest négociants to expand and modernize their production facilities. One of the more innovative examples of this trend can be seen in the firm of Veuve Pommery. In order to accommodate the need for larger production facilities, the famous widow took a chance on moving both her main residence and her business premises to a piece of land at the edge of the city of Reims that was said to be "ill-suited for construction."[116] Attracted by the immense underground chalk galleries, which proved ideal for wine cellars, Pommery began building in 1868. Work was completed on her immense project in 1878, and it included her primary residence, all production facilities, cellars, and showrooms, along with an immense garden laid out along the lines of those of seventeenth-century British aristocratic country houses. Here she would be at the center of producing and marketing the wine that carried her name around the world. Some commentators were impressed by the way the chosen architectural design evoked the *tradition bordelaise*.

Pommery's building scheme worked to her advantage by further associating her brand name symbolically with tradition and quality. Enormous status was invested in a firm's brand name by the late nineteenth century, and firms dissociated themselves from mass production. Pommery's new facility, recalling the traditions of Bordeaux, could offer consumers a sense of continuity with quality wine production and old-world traditions. A number of travel books and promotional pamphlets prominently featured drawings of the new Pommery château and production facilities.[117] These pamphlets were aimed specifically at the British middle classes, who were avid consumers of Bordeaux wines, once a symbol of the rich and powerful of eighteenth-century England.

While some of the smaller producers created production facilities adjacent to family compounds, the largest producers after mid-century increasingly chose to build their primary residences away from the center of production. Indeed, some of the most prestigious négociant families moved to estates with vineyards in the countryside surrounding the main manufacturing centers. Women like the daughters of Veuve Clicquot and Veuve Pommery made this their main home, removing them from the day-to-day operations and commercial activities of the firm. Statistics show that négociants owned almost 3,400 hectares of vines in 1893. It remained difficult, however, to amass enough small parcels to create a sizable contiguous holding. In the town of Tréloup, for example, the 2,099 hectares of vine land reported in 1899 were divided into 33,000 individual parcels.[118] Contemporary sources, including accounts by visitors at the end of the century, indicate that there were a number of prominent négociants who did not purchase land because they "objected to being embarrassed with the cultivation of such tiny plots."[119]

Reports by the U.S. consul at Reims suggest that regional landholding by négociants was increasing throughout the last quarter of the nineteenth century. "The heads of the great champagne houses are as a rule proprietors, in one or more locations, of extensive vineyards," he wrote. "They possess also," he continued, "a large winemaking establishment and retain a regular troop of vine dressers whom they lodge and employ by the year."[120] Mumm, established in 1827 by the sons of Peter Arnold Mumm, a vineyard owner from the Rhine region of Germany; Friedrich Giesler, also of Germany; and the French négociant G. Heuser typified this trend. The Mumm family gradually built up the reputation of the firm, particularly with consumers abroad. Although the firm owned no vines by the 1850s, it began to purchase grapes directly from growers for processing in a new family-owned wine press near Verzenay. The firm made its first vineyard purchase in 1882, buying eight hectares of the most expensive land in the Côte des blancs. Slowly, throughout the 1880s and 1890s, as mildew and phylloxera took their toll on small growers and land became available, the firm amassed a considerable vineyard and a sizable group of employees. By the time that the head of the firm, Hermann Mumm, who retained his German citizenship, was put under house arrest with other non-French professionals as a potential danger to the nation in 1914, the family owned over fifty hectares of the finest grape-producing land.

Perhaps more important than landholding was the tendency of the largest négociants to establish what the U.S. consul called "large winemaking establishments"—grape presses—in the vineyards. The vignerons' grapes were taken straight to the presses of "the great champagne houses—Moët et Chandon, Clicquot, G. H. Mumm, Roederer, Deutz et Geldermann, and others—all teeming with bustle and excitement, and with the vines almost reaching their very doors," Henry Vizetelly wrote.[121] Direct sales of grapes to négociants had been a long-established practice in the montagne de Reims, but by mid-century, as the case of the Mumm firm confirms, the practice had become more generalized. Wine presses were built by négociants, particularly in areas where they were most interested in purchasing grapes. The négociants could, in theory, create a better product by supervising the wine from pressing to first fermentation, freeing the vignerons from "the concerns of this delicate procedure."[122]

The practice of buying vineyards and installing grape presses owned by the champagne firms gave négociants a powerful presence in rural affairs. In particular, it served to link négociants with local landholding elites. By the 1880s, for example, négociant families that owned vineyards and large landholders around Épernay joined to create the Syndicat viticole d'Épernay. Grouping 120 large landowners, the syndicate created an influential organization to control the wages of *ouvriers viticoles*, often smallholding vignerons. Its leadership came from family members of négociant families, including the vice president, Gaston Chandon. Economically and socially powerful, the syndicate refused to respond to questions posed by the government regarding its goals and membership. The prefect of the Marne reported that the organization included "people who are notoriously hostile to the government" and "could be a weapon of reaction."[123] It was not simply négociants in the Syndicat viticole d'Épernay who appeared "hostile to the government." Wine interests, which had profited from political connections to the aristocracy and the ruling elite of the Second Empire, continued to support right-wing candidates and political causes well into the twentieth century.[124] Négociants were generally viewed by supporters of the republic as reactionary. Long after the right-wing causes they supported had subsided as imminent threats to the republic, the négociants were seen as anti-republican.

Négociants were not just royalists; increasingly, they were aristocrats. Most champagne families had rather modest, bourgeois origins, but during the nineteenth century, as commentators wryly noted, more and more of them acquired titles of nobility, mainly through marriages between the daughters of grandes marques négociants and the sons of old aristocratic families.[125] These titled family members often lived on the family estate, overseeing the cultivation of family-owned vineyards or the operations of the firm's grape presses. Marriage into a champagne industry family was, in many ways, an ideal source of income for nobles, because "manufacturing initiatives that combined familial resources with seignorial responsibilities were eminently respectable."[126] For the négociant families, who willingly exchanged their personal wealth for status, obtaining these alliances linked the champagne commercial elite with aristocratic tradition and prestige. Champagne brands were often family names, and "strenuous efforts were made to give them circulation."[127] Noble titles and family coats of arms could be added to the firm's name or logo, giving a certain air of distinction and prestige to the wine.

Aristocratic links were beneficial economically, but politically and socially, they allied the négociants with an increasingly marginal elite. There are a number of possible reasons why négociants chose to pursue these alliances. One was the gulf that separated the champagne négociants from other commercial elites, particularly their counterparts in textiles. In 1876, the U.S. consul noted that the region was marked by long-standing "feuds and jealousies which exist here between wine and wool."[128] "Feuds" between wine and wool had their roots in political differences. While the regional commercial elite, particularly in the textile industry, joined with the *nouvelles couches provinciales*—the small and mid-level bourgeoisie of the professions and small businessmen who were mildly liberal—to support the conservative republic, the champagne négociants generally remained Bonapartists or royalists.[129] Given the political tendencies of the négociants, links with aristocrats who shared their political convictions kept both groups from political isolation.

Whereas feuds between wine and wool were political, the jealousies were economic and social in origin. For the traditional woolen families, the champagne négociants were little more than parvenus. Links to old French families gave négociant firms a legacy while also helping to mask the predominance of foreign names among the region's most prestigious wine houses, and the grandes marques in particular, which attracted considerable attention from supporters and critics of the industry alike. Competitors and local elite were especially rankled by the success of champagne makers with German names, such as Giesler, Heidsieck, Koch, Kunkelmann, Mumm, and Walbaum, many of whom were former clerks or salesmen.

Rivalries and suspicions between immigrant négociants and their French counterparts appear to have waned after mid-century. Historians have often attributed this to these marital alliances and the rapid integration of the foreigners into the champagne collectivity. To this could be added a series of legal and economic challenges to the collective interests of the négociants that undoubtedly also served to form bonds of solidarity and obscure divisions. Of growing importance to the industry as a whole was the long and very costly legal battle in the French courts for the protection of brand names and the exclusive use by regional négociants of the denomination "champagne." Individual négociants undertook lawsuits or circulated petitions for changes in the laws affecting the industry, but, increasingly after mid-century, négociants joined together to use their combined financial resources to pursue the lengthy legal procedures against the use of the appellation "champagne" by producers outside the region.

Given the relatively recent emergence of wine production as an industry, there was no body of law for the protection of regional denominations. Existing laws presented numerous problems when applied to the champagne industry. To bring lawsuits against manufacturers of sparkling wines outside of the region who used the regional name of "champagne" as a generic label, the champagne négociants had to prove that they were in fact manufacturers or producers of champagne. The description "négociant en vin de Champagne" implied that they were engaged in commerce, simply selling wines, not in production of the finished product. To some extent, the problem with the professional title was eliminated by the rapid development of mechanized production facilities, which gave champagne houses the air of "factories." The négociants and their modern "wine factories" met expectations of what constituted an industry. Nonetheless, the négociants spent considerable time in court until the 1880s proving that they had a right to legal recourse as manufacturers.

An equally serious problem for the négociants was proving to the courts that champagne fit the definition of a manufactured product, and that the name "champagne" was thus industrial property, protected under the law. French courts recognized that the wine called champagne came from a unique historical region and ruled that the still wines gave the sparkling beverage its unique character. The appellation "champagne," it was therefore determined, specifically designated wine from the Champagne region that had been rendered "sparkling" (*mousseux*) in a second processing phase. Although the process had become highly industrial, the name "champagne" was not generic. Négociants from Champagne thus had the exclusive right to use it on their product.

Collective efforts for control of the regional denomination of champagne served to forge links among the négociants. Along with these shared economic concerns, a series of conflicts within the regional Chamber of Commerce also served to unite négociants. Throughout most of the nineteenth century, the Chamber of Commerce regularly reported on issues regarding champagne production, but it rarely acted to represent the collective interests of the industry. In an organization dominated by textile magnates, issues concerning the textile industry were the central concern, and négociants grumbled that the concerns of champagne were marginalized. Frustration mounted in the 1870s, when the champagne industry, in the midst of an economic slump because of the disruption of foreign and domestic markets by the Franco-Prussian War, received no additional attention from the Chamber of Commerce.[130]

Contention between the two groups culminated in 1882, when the Chamber of Commerce released an "official" document regarding the decline of champagne sales to the United States. According to this report, the sales slump was not caused by competition from sparkling wine producers in California, as the négociants had maintained, but by "quality problems" created by regional champagne firms that shipped poor-quality champagnes abroad.[131] In the midst of legal action for the protection of the regional denomination, négociants expressed outrage at the untimely accusations by the Chamber of Commerce. Led by Florens Walbaum and representatives of the firms of Mumm and Giesler, all major champagne exporters, the négociants publicly denounced the report, angrily demanding a retraction and additional representation of négociants in the Chamber of Commerce.[132]

When the négociants' demands were met publicly with silence, twenty-nine of the grandes marques came together to create a representative body of their own, the Syndicat du commerce des vins de Champagne. Membership in the Syndicat was extended only to a small group of firms. Between 1882 and 1900, it never included more then fifty-four firms, all of them grandes marques. If there was any lingering resentment within négociant circles of foreign-born négociants, it was not apparent in the organization of the Syndicat. Nearly one-third of its membership in 1895 were first and second-generation immigrants. While the Syndicat was, theoretically, open to all "négociants or former négociants of wines from the department of the Marne," the grandes marques never invited the small-scale producers to participate.[133]

Despite its restricted and rather exclusive membership, the Syndicat du commerce created the impression of a powerful, unified négociant presence in local as well as national affairs. With the considerable economic power and national and international recognition of its membership, this professional organization was generally acknowledged by those within the industry, as well as by the French state, as the representative body of the champagne houses and négociants. Under its auspices, négociants acquired a collective "voice," discernible in public debate. Publicly, there was little opposition to the grandes marques assuming the leadership role in the industry. Nevertheless, as we shall see later, conflicts among the négociants eventually led to the formation in 1911 of a second organization, the Syndicat des négociants en vin de Champagne, so that there were two competing voices. But until the riots of 1911, few questioned the right of the Syndicat du commerce to represent négociants' interests.

The establishment of the Syndicat marked not only the emergence of the grandes marques firms as leaders of the champagne industry but also the emergence of the Champagne négociants as a united, coherent group within the commercial elite of the region. Using the considerable financial resources of its membership, the Syndicat pursued collective legal action and lobbied the government for "the protection, both in France and abroad, of the trade in *vins mousseux* from Champagne."[134] A strong central organization was important in the late nineteenth century, when négociants found their domination of the sparkling wine market challenged. "The wine of Champagne has suffered the fate of all great discoveries," wrote one négociant, "a multitude of imitations rushed forward, and the leprosy of counterfeits clung to its vogue."[135] The Syndicat du commerce engaged in a legal and legislative battle to protect the regional appellation from being used as a generic label for sparkling wines and brand names from imitation or counterfeit production. The appellation "champagne" was not yet protected by law, and wines originating outside of the region frequently used the name. Calling their products "champagne," competitors often adopted the style of champagne labels and packaging.[136] "Champagne," the syndicate argued in legal action, could only properly be used by the community of producers within the confines of Marne.[137]

As a luxury product actively associated with the old-world quality of France, champagne was a highly visible symbol for foreign governments to tax. By 1883, reports in the *Vigneron champenois* estimated there were over thirty different sets of regulations and tariffs that applied to sparkling wines.[138] A bottle of champagne that sold for an average of 5 francs in France at the end of the century, would have an additional tax of 3.5 francs added in the United States and about 4.76 francs in Russia, almost doubling the price of the wine, and champagne was slapped with higher tariffs during the frequent trade wars of the day. However, by 1890, the growth of domestic markets for regional sparkling wines had given the industry new vitality.

While foreign markets and the grandes maisons continued to be central to the economic strength of the region, the importance of smaller firms to regional economic prosperity grew in the 1890s. After a period of slow growth in sales between the 1850s and 1870s and a dramatic fall in sales during the economic crisis of the 1880s, domestic sales reached record levels in the 1890s, soaring from 1,176,189 bottles in 1890 to 6,680,923 bottles in 1900. Small producers catering to the domestic market were increasingly important to the regional economy. Although there was a narrow market for the brand-name champagnes of the grandes maisons, French consumers showed a marked preference for the cheaper *vins de deuxième choix*. These firms captured the French market by building on the symbolism and meanings transmitted and transformed by brand-name champagne firms. The promotion of the regional appellation of "champagne" was particularly effective with consumers, as witnessed in the many wine labels destined for the domestic market. These labels, by the turn of the century, were more likely to feature symbols of modernity—bicycles, sporting events, airplanes, the French Revolution—than the brand names and symbols of gentility so appealing to foreign consumers (figs. 14 and 15). Champagne was portrayed less as an aspect of a timeless France than as one of modern France. Luxury, however, remained a constant. Within the French market, these firms catered to consumers' desire for a fashionable product with a range of choices of varying quality at relatively low prices.

The surge in markets for vins de deuxième choix increased the visibility of "champagne" among French consumers, while also enhancing the economic and social power of some of the small regional producers. The Syndicat du commerce, still primarily focused on external markets, made it clear, however, that it would set the limits of acceptable practices among the négociants. In particular, it publicly required that all wines bearing the name "champagne" be from grapes harvested (*récoltés*) in Champagne and that all aspects of production (*manutention*) be completed in the region. All négociants were expected to conform. When the firm of Mercier refused to stop shipping regional still wines to cellars in Germany and Luxembourg, where they underwent a second fermentation and became "champagne" without the burdens of export taxes, the firm was expelled from the Syndicat and excluded from négociant social affairs.[139] The expulsion had little effect. The case of Mercier was never discussed in the press, and the only dialogue between Mercier and the Syndicat was in internal Syndicat documents. The Syndicat brought legal action against German firms who committed exactly the same kind of "fraud" as Mercier. When the same abuse involved a négociant within the region, the Syndicat dealt with the issue outside of public scrutiny; the problem with German "champagne" production, by contrast, was widely recognized within the wine industry and discussed in the press.

The enormous amount of labor and capital needed to create the sparkling wines of Champagne, whether destined for foreign or domestic markets, made wine production there fundamentally different from that of other wine regions. For most wine regions, wealth was in the land, and markets were controlled by the major landholders. The main costs of producing wine, even in fine wine regions, were in the cellar and the casks. Champagne was different. Not only were regional still wines generally not profitable, but, in the end, they were indistinguishable in the unique blended, sparkling finished product of each firm. The individual attributes of the terroir—so highly prized in Bordeaux, for example—would be difficult, if not impossible, to distinguish even by a highly educated consumer. Additions of sugars and liqueurs further distanced the finished product from the original grapes and terroir. Moreover, négociants were under no legal obligation to use only grapes produced in the region. Court rulings had stressed the place of manufacture and, to a lesser extent, the origins of the wines. Judicial decisions were vague about boundaries to terroir. The decision by négociants to reduce the use of names of villages, such as Sillery and Aÿ, in favor of the label "champagne" reflected not only the industrial nature of the finished product but also the growing power of the manufacturers to control notions of origin. With the highly mechanized transformation of the wine and the complex blending formulas distancing the product from the land, the reputation and power of the manufacturer, as René Lamarre stressed, became everything.

❦

The interdependence of the rural and the nonrural world, the world of peasant grape growers and industrial manufacturers of champagne, was brutally demonstrated when a worldwide economic slump in the 1880s checked the steady rise in sales. Although France was relatively well off during the depression, champagne exports declined by almost 20 percent, and the domestic market dropped by 15 percent. After record high prices in 1880, base prices for both white and red grapes plummeted. Poor weather conditions in 1883 reduced grape supplies and brought a slight jump in prices, but many vignerons had little to sell.[140] In 1885 (and again in 1892–93), there was a severe drop in sales. Prices reached their lowest point in two decades, and only one-third of the grapes harvested in 1885 were purchased by the négociants.

The burden of the economic slump of the 1880s seemed to fall disproportionably on the vignerons. With over 300,000 hectoliters of wine in storage, 80 million bottles of champagne in cellars, and only 17 million bottles of champagne sold to consumers in 1885, regional négociants had little incentive to purchase grapes.[141] The vignerons had few outlets for their production; grape prices were too high for ordinary wine production, and the vignerons had long since abandoned their wine presses. As one vigneron explained to René Lamarre, "Young people did not buy presses, old people sold them or left them in disrepair; apart from a prudent few, the others no longer have vats, barrels, absolutely nothing . . . nothing that is necessary, and above all no buyers."[142] Despite temporary reductions in Midi production caused by the spread of phylloxera,[143] even the "prudent few" still able to produce red wines reported a substantial drop in revenue.[144]

The economic slump of the 1880s produced "a gradually increasing depression in the Champagne wine trade in all countries."[145] Peasant producers, who had profited from their relationship to the champagne manufacturers during the years of general prosperity, now argued that the négociants were exploiting the rural population. The creation of the Syndicat du commerce in 1882 fueled these accusations. "The Syndicat," a writer for *Vigneron champenois* declared, "has joined together to fix prices."[146] In the minds of the struggling vignerons, the formation of the Syndicat, which united the most important purchasers of local grapes and some of the largest landholding families of the region, seemed inextricably linked to their own hardships.

Tensions grew with increased activity in vineyard affairs on the part of some of the more prominent négociants and their families. Although négociants owned wine presses and property in rural communes, they had largely remained aloof from affairs in the vineyards. During a serious infestation of harmful cochylis moths in the mid 1880s, several members from among the grandes maisons experimented with mechanized treatment methods. Négociants who had large landholdings in infected areas donated the machines that they had selected to local villages. Although this was seemingly a charitable gesture, they then proceeded to lobby local officials to force vignerons to use them.[147] Attempts by officials to force peasant compliance created anger and a *vive agitation* of the rural population well into the 1890s.[148] The négociants appeared not only to be fixing prices but also to be meddling with the vignerons' sacred link to their terroir.

In 1889, after nearly ten years of consistently low prices, a dramatic surge in grape prices, exceeding those paid during years with similar quality and quantity, was greeted as the end of the years of hardship. Syndicat du commerce records show that prices paid for grapes nearly doubled over the previous years. Prices for red grapes in Bouzy, for example, reached up to 1,400 francs *la pièce*, and white grapes from Avize reached 1,200 francs *la pièce*.[149] In the commune of Aÿ, the average price of a hectoliter of grapes was 155 francs in 1886 and 1887; prices jumped to 220 francs per hectoliter in 1888 (although with an average production of only 7 hectoliters per hectare, there was little to sell); the year 1889 saw an astronomical 531 francs per hectoliter.[150] For the first time in nearly a decade, the quality of the harvest appeared to correspond to the prices offered by merchant-manufacturers. But elevated prices were the result of an unusual bidding war between rival firms interested in creating a *vin du centenaire* for the anniversary of the French Revolution.[151] Without the rivalry of firms in 1890, prices for the harvest—which was only slightly lower in both quality and quantity—plummeted by over 50 percent.[152]

By the end of the harvest of 1890, the idea that "all is lost, save honor" had a particular resonance with the rural population. Amid growing vigneron discontent, René Lamarre distributed his pamphlet *La Révolution champenoise*. Unaffiliated with any political party or organization, Lamarre, in both his pamphlet and subsequent meetings in vine-growing communes, appealed to the Champenois vignerons to recapture their economic legacy— *notre propriété collective*—the name "champagne."[153] Using language very similar to that of French socialists such as Jean Jaurès, the nineteen-year-old vigneron laid out his program for restructuring the economy of Champagne. All vignerons were to join in a central cooperative. Over the course of three years, they would make wine, receiving payments for their grapes based on the average price of the six most recent harvests. Funds during these early years would come from a loan guaranteed by the collective value of the vineyards. During the third year,

Lamarre projected, the vignerons would begin to produce their first bottles of champagne, which they would market under a collective label. By this point, according to Lamarre, the majority of négociants would have empty cellars and no grapes for restocking. Unable to supply their clients with true champagne, the négociants would be forced to capitulate to the power of the vignerons.[154]

Because the vigneron-based organization would sell "*only* products of their harvests," he argued, consumers could be sure of the purity, the quality of regional sparkling wines. "You alone will have the right to use the name: Champagne," Lamarre continued.[155] He projected that demand would be high, but supplies from the limited area of "true champagne" would remain stable. For every hundred consumers, he estimated, only a privileged twenty would be able to purchase regional sparkling wine. This he illustrated with an imaginary dialogue between a waiter and a customer at a fashionable restaurant. "Waiter, some *Champagne!*" the customer demands. "At 40 or at 50 francs [a bottle]?" the waiter asks. When the customer hesitates in disbelief, the waiter continues: "We have some Chablis at 6, 8, and 10 francs [a bottle]." Big spenders like the imaginary customer would respond with, "Give us some *Champagne*," according to Lamarre. They could do no less, because "the true [champagne] is never too expensive."[156]

With this monopoly, riches, even opulence, were lyrically promised to all Champenois: "Think it over carefully everyone—merchants, workers, artisans; when the 140 million in profit that is earned each year is shared among 18,000 families, instead of remaining in a few hundred hands, when those who are poorest today are the richest tomorrow, everyone, even the most humble, will feel the effects. The vigneron, rather extravagant and generous by nature, will provide not only well-being and comfort, but abundance." The vignerons of Champagne *viticole* would be the vanguard of prosperity, funneling the riches of champagne back to the rural community. The négociants, who he maintained were Germans, were not to be included in the community of Champenois. In thinly veiled references to the loss of Alsace-Lorraine in 1870, he declared these German négociants to be *voleurs* (thieves), stealing the property of the patrie.[157]

Critics ridiculed Lamarre's self-declared "honest and profitable monopoly." Regional négociants decried the "audacious propagandist," and enlisted the help of a number of *bons propriétaires* to spy on his activities.[158]

The economist Charles Gide, in an eloquent critique of Lamarre's project, wrote:

> Cooperation has never been the exploitation of consumers by producers. Certainly, we would be delighted to think that the money spent on champagne would go into the pockets of the petits vignerons rather than into those of the large merchants, but we would become obligated to pay to these victims 25 francs a bottle, as M. Lamarre graciously announced. Oh! Oh! in that case, long live Moët and Chandon! Long live even the duchesse d'Uzès! We shall continue to drink their wine; we shall even drink to their health.[159]

Even socialists based in the town of Épernay, who actively tried to recruit Lamarre to join the Parti ouvrier français, did not openly support his proposal, being opposed to his activity in the vineyards before the workers in the main regional industrial centers were organized. By the late 1890s, he was completely at odds with Guesdist socialists in Épernay, rejecting both their internationalism and their support of Alfred Dreyfus in the political cause célèbre of the Third Republic. Censored by the POF for declaring the defense of Dreyfus an "abominable struggle against the patrie," Lamarre definitively broke ties with the socialists, running as a "candidate for the vignerons" against the POF candidate, Paul Vallé, in the legislative elections of 1898.[160]

But while critics scorned Lamarre's "utopian" project, he was successfully capturing the discontent of the vignerons.[161] Only a few thousand vignerons rallied to Lamarre's proposed cooperative, producing "Champagne des Trois-huit," sold in Paris consumer cooperatives.[162] Thousands more, however, attended his meetings. According to one paper, Lamarre emerged as "one of the masters of Champagne."[163] Through his meetings and later editions of *La Révolution champenoise*, he helped to formulate a sense of solidarity among vignerons in Champagne *viticole*—a sense of solidarity based on the idea that the economic legacy of champagne was the exclusive property of the rural community, serving as a crucial element in the transmission of the qualities of its terroir.

Lamarre focused attention on the relationship between the vignerons and négociants. Throughout his pamphlet, he carefully avoided mentioning négociants with decidedly French names, preferring to cite names like Roederer, Planckaert, and Bollinger. His anti-German rhetoric linked *revanchisme*—national revenge for the loss of the Franco-Prussian War—to the recapturing of "champagne" from the hands of the German négociants. Nationalism merged with class struggle in Lamarre's formula, creating a certain populist appeal. "Monopolizing the name champagne" became a struggle for protection of the terroir, the patrie, and the rural community.

"The négociants," Lamarre bellowed to a throng of vignerons, "are all Germans who came to France without a *sou* . . . they do nothing; however, today they are millionaires with châteaux, while you, the proprietors, you work, and you are unhappy, it is your work that enriches your exploiters."[164] Enthusiastic cheers came from the crowd of vignerons, who believed that they were the center of the community, the backbone of champagne production. "Tout est perdu, fors l'honneur," Lamarre proclaimed. For vignerons throughout the Marne, faced with falling incomes and anxiety over reports of a mysterious new vine disease called phylloxera, the négociants and the Syndicat du commerce were usurping their property, compromising the collective interests—the terroir—of the champagne community. Protecting the French "national fortune" meant strengthening the connection of France and champagne by promoting the link between sparkling champagne and the terroir. All did, indeed, seem lost, except for honor.

CHAPTER

RESISTANCE AND IDENTITY

Cultivation Methods and
the Wine Community, 1890–1900

Social tensions in Champagne were captured in a cartoon that appeared in regional newspapers in the early 1890s (fig. 16). Standing on the edge of a vineyard, a vigneron is unsheathing a sword. "I really think that's it, the so-called phylloxera," he says, gazing across the vineyard, not at the minute plant louse that had been slowly destroying France's vineyards since the 1870s, but at a well-dressed bourgeois eating the grapes.[1]

Phylloxera's arrival in the region focused attention on the precarious situation of those who supplied the raw materials for champagne production and the gravity of the social conflicts brewing in the champagne industry in the 1880s. The spread of phylloxera and the subsequent drop in grape supplies presented a serious challenge to the industry as a whole. But it was ultimately the conflict over the creation and control of the quasi-interprofessional syndicate formed to deal with treatment of the vine blight in the 1890s that served to expose the important fissures within the Champagne community. For the small proprietors and growing rural proletariat, the "dreaded parasite" that roused them into violent confrontation with the authorities was not the tiny pest of the vine but the bourgeois outsider so skillfully portrayed several years earlier by René

Lamarre, the négociants and large landowners connected with the grandes marques and markets far removed from the terroir.

Historians have generally concluded that the conflict over phylloxera in Champagne was the result of peasant ignorance or fierce rural individualism. In this interpretation, peasant grape growers either naively refused to believe that the bug could infest their vineyards or selfishly resisted collective efforts to treat the blight. In the face of the infestation, the vignerons seemed to conform to the classic Marxist portrait of peasants more active in a primitive "exchange with nature than in intercourse with society."[2] These conclusions are backed by evidence from the most widely examined sources documenting the arrival of phylloxera in the Marne. In these sources, négociants and bourgeois observers passionately enumerate the errors of the peasantry in the struggle over the creation, composition, and control of the Association syndicale autorisée pour la défense des vignes contre le phylloxéra (hereafter the Association syndicale). The picture is one of backward peasants opposing well-meaning, forward-looking négociants and large landowners who sought only to preserve Champagne's economic heritage in the face of the blight.

The drawing of the angry peasant identifying the bourgeois as the real vineyard parasite suggests that more was at stake in the minds of the vignerons than simply a debate over vine treatment methods and syndicat organization. Moreover, if the events in Champagne are viewed in the context of France's overall battle against phylloxera, the picture changes dramatically. By the time phylloxera was firmly installed in Champagne, scientists and savants had hardly come to any firm conclusions about the most effective treatment methods. Debate raged between those who saw the tiny aphids as the source of the problem and those who believed that the bugs were a secondary effect of a more serious contagion. Peasants who took similar albeit less well articulated positions, based on vineyard experience, were, however, deemed ignorant. One historian has argued that the growers denied that aphids were the source of the problem and that "their readiness to seize on the most flimsy evidence to support this conviction suggests a deeper psychological base for their incredulity."[3] By the late 1880s, "backward" peasants, unaided by a scientific community, which proposed everything from spraying goat urine on the vines to burying frogs near infected vine roots, had taken matters into their own hands. Throughout rural France, peasant vignerons experimented with grafting native grape scions onto resistant strains of American rootstock. While the decision was not based on scientific method, it was, ultimately, a rational one.

An examination in this context of peasant newspapers, petitions, and reports, as well as unexplored archives of regional négociants, results in a more nuanced picture of the turbulent events that surrounded the arrival of phylloxera in the Marne and the formation of the Association syndicale. The contention over the phylloxera was not simply the result of the "obstinacy of the peasants" or a latter-day jacquerie.[4] Peasants in Champagne, like peasants throughout France in the early 1890s, resisted chemical treatments on the grounds that they were both costly and ineffective. This was based on experience, not ignorance. Some vignerons in Champagne did, indeed, doubt the existence of phylloxera. But understanding this does not involve a search for "a deeper psychological base" for this attitude. Indeed, this position was not far removed from that of many contemporary scientists who, despite having seen the aphids, resisted the idea that they were the source of the vine disease.

Broader economic changes in Champagne had altered the social fabric and economic relations of the wine community in the years leading up to the phylloxera crisis. With the phylloxera battle portrayed as no more than an isolated incident, however, the history of the bitter struggle for control of Champagne's economic and cultural legacy in a restructured marketplace is masked. By the time phylloxera made its entrance into the Marne, vignerons had become entirely dependent on the négociants for the sale of their grapes. There was, however, no guarantee—regardless of the price of a bottle of champagne—that growing wine grapes for sale would be profitable. Recent economic hardships in the 1880s had shown that the vignerons bore the brunt of radical fluctuations in the wine market. The vignerons were at a disadvantage, and they knew it. They were not without recourse, however, as Lamarre had so recently brought to their attention. Terroir, their crucial connection with the final product, was still, despite accelerating shifts in landownership, under peasant control.

Phylloxera threatened to destroy the vines, putting both the vignerons' livelihood and the négociants' reputations at risk. It not only compromised the regional grape supply but disrupted production of all French wine, the nation's second most important export—the source of 25 percent of the country's annual agricultural revenues. The effect on electoral support for the Third Republic was profound. Decisions about how to preserve the vines were thus more than simple matters of local politics and economic self-interest. Champagne had become a national good, associated with France's economic health and cultural heritage.

The struggles of the phylloxera era saw attempts by both vignerons and négociants to renegotiate the community centered on terroir. Each side looked to the state—the vignerons to the Ministry of Agriculture, the négociants to the Ministry of Commerce—to support its program for protecting the cultural legacy of Champagne by emphasizing its connection to the larger ideal of France. Although locked into a shared community of interests, these groups differed not only over control of the terroir but about its very meaning.

❦

When the first signs of the phylloxera were detected in Champagne in the late 1880s, the vine blight had already caused considerable destruction and economic dislocation in the south of France. Experts estimated that what they called *Phylloxera vastrix* (today *Dactylasphaera vitifoliae*), a minute insect that attacks the roots of vines, was introduced in Europe sometime in the 1860s from the United States. The phylloxera aphids, although unidentified, had long been destroying American vineyards planted with European varietals. Because of the popular belief that vines, particularly European varietals, were not suited to the soil or climate of North America, few American vine growers researched the source of their vineyard problems. Although the transport of vine cuttings, rootstock, and seeds across the Atlantic was common practice before the 1860s, the long sea voyage killed phylloxera aphids before they reached Europe. Advances in transportation—the introduction of steamships for transatlantic travel and the spread of railroads—not only greatly expanded the market for French wine, particularly champagne, but also promoted the survival of the aphid that would eventually menace the French wine industry.

Phylloxera did not "appear everywhere at once," one historian notes, "and its impact was variable in time and space."[5] The aphids destroyed vineyards slowly, literally sucking the life out of the vines and the wine-producing communities. The *nouvelle maladie*, as it was initially called, was first investigated by a non-government-sponsored organization in the Hérault in the south of France as early as 1869. Led by Jules-Émile Planchon, a physician and professor of pharmacy at Montpellier, Felix Sahut, a well-respected vinegrower and author, and Gaston Bazille, a large property owner in the Hérault, the team quickly discovered the infestation of aphids on the roots of anemic vines. The results of their investigation were widely published in the local and national press.[6] "It is useless to look elsewhere for the cause, which is unhappily all too evident," a contemporary observer wrote after the Planchon announcement; "what is now necessary is to find the remedy."⁷ Finding the remedy for most plant diseases or insect attacks had previously been a simple practical matter that did not require much technology or scientific investigation. Phylloxera proved to be different.

Leading scientists throughout France scoffed at the findings of the provincial "entomologists," as they called them (this was not their profession).⁸ Scientific resistance from the professional community, particularly in Paris, mounted. Despite clear visual evidence from southern France that the vine blight was the result of aphid infestation and damage, many of France's leading scientists maintained that the aphid was only a secondary effect of poorly tended vines. Far from the vineyards, botanists like Charles Naudin, the director of the famous Jardin des Plantes in Paris, argued forcefully that the vines, all in areas of ordinary wine production, had succumbed to a mysterious disease, which, ultimately, made them vulnerable to the plant lice identified by the Planchon team. Aphids themselves were thus only a secondary presence following the spread of disease. Some, following this line of thinking, even argued that the presence of the physiological disease was connected with a special local terroir and poorly adapted cultivation methods.⁹ Given that these same French scientists had long been interested in the influence of local environment and terroir on the body, it is perhaps not surprising that when faced with a mysterious vine malady, they sought a connection with France's unique terroir.

The intense controversy drew in some of France's most prominent names in botanical science. All levels of the academic community, from the prestigious Académie des sciences to schools of agriculture, entered the dispute over the nature and etiology of the vine disease. While scholarly debate raged, growers, local landowners, and agricultural workers moved away from theory and began experimenting with practical solutions. Various trials of eradication were conducted: flooding vineyards; using substances with insecticidal effects; and planting with American vines, some of which were mysteriously resistant to aphid attacks. The government of the Third Republic also stepped into the fray, creating a Commission supérieur du phylloxéra under the direction of the minister of agriculture. The end result of over a decade of wrangling was the discovery of the phylloxera paradox, as Planchon first termed it: the source of the disease—the origin of the aphids—and the cure were the same: American vines.

The issue of possible replanting or grafting of French with American vines was not a simple one. Not all American vines were equally resistant to phylloxera; not all were equally suited to the various French terroirs; and not all when grafted with French scions produced acceptable wines. Early experiments with American plants, mainly conducted in labs away from the highly prized terroir in order to avoid the transmittal of new infestations, were a disaster. Under pressure, particularly from quality wine producers who were skeptical about American vines, and with the financial support of the French government, many of France's most distinguished scientists turned their attention to finding some chemical or biological entity that could kill the aphids and save the surviving French vineyards. Hundreds of possibilities—from folk remedies to chemical compounds—were tried. Responses remained highly eclectic, in no small part because of a persistent uncertainty about how geology, climate, and the elusive terroir affected the bug and its eradication.[10]

Numerous conferences, such as the famous Bordeaux Conference of 1881, one of the most important gatherings of scientists and savants interested in the phylloxera question, rarely led to consensus. The Bordeaux Conference, for example, split definitively between two distinct factions: the "chemists," who, following the lead of the Académie des sciences, saw phylloxera as the result of disease and promoted carbon bisulfide treatments and chemicals, and the "Americanists," who, following the 1873 work of the Franco-American team of Planchon and C. V. Riley, believed varietals of resistant American vine roots could be used in grafting with French scions for mass replanting of vineyards. Decades of constant debate and experimentation would not result in a definitive technique for eradicating the pest. Much to the dismay of quality producers, grafting appeared to be the only solution, but it remained highly controversial. After all, many would have agreed with the famous French chef and gastronome Auguste Escoffier that "French soil enjoys the privilege of producing naturally and in abundance the best vegetables, the best fruits, the best wines in the world."[11] France's national greatness had to be defended from the ground up.

Phylloxera made its way to northern French vineyards in 1878, although it did not immediately appear in the department of the Marne. As we saw in Chapter 3, Georges Vimont, who would later emerge as a leader among the small proprietors, sought to form a peasant-based organization to study phylloxera and possible treatment methods, but the departmental administration instead created the Comité d'étude et de vigilance contre le phylloxéra (Vigilance Committee), which consisted mainly of agricultural professors, large landowners, and négocians. On the appointed twenty-eight-member central committee that directed the work of the organization, there were only three vignerons.

After several trips to the Midi in the 1880s to observe the progress of the disease, the Vigilance Committee, along with members of the Syndicat du commerce and the Société d'agriculture, de commerce, des sciences et des arts de la Marne, reiterated an opinion popular in other agricultural regions: phylloxera was not the source of the vine problem but an effect of poorly cultivated vines. "Many members of the committee," one contemporary argued, "did not believe in phylloxera."[12] They saw little reason to fear that phylloxera would attack the healthy vines of the exceptional terroir of Champagne.[13] In a meeting of the Syndicat du commerce regarding the phylloxera invasion of the neighboring department of Seine-et-Marne in 1886, for example, a wine specialist from the Vigilance Committee argued that "the considerable differences that exist between the cultivation, planting, and soil of the vines of Champagne and the vines of the Seine-et-Marne can perhaps preserve us from the invasion of the blight."[14]

Others were not so optimistic. Vimont, despite having been rebuffed, continued his efforts to create a peasant-based organization. The phylloxera, he argued, would come regardless of terroir.[15] In April 1881, he made a passionate appeal to the Comice agricole et viticole d'Épernay, a small yet powerful group of individuals interested in agricultural issues in and around Épernay, to create a series of independent communal associations, or *syndicats*. His project involved creating such syndicats in each commune to regularly survey vineyards for signs of infestation, follow current research on phylloxera, study other diseases of the vine, and experiment with treatments. Signatures from local vignerons interested in participating were presented along with a proposed adherence form. Vimont envisioned asking for state money to support the efforts, but he stressed the importance of keeping all such organizations autonomous, small-scale, and in the hands of the vignerons. He argued that since Champagne was not yet under direct assault from the bug, the syndicats should be primarily involved with educational efforts. When the time came, the syndicats could be transformed into treatment organizations.

Vimont's proposal was met with great skepticism. Many argued that the plan simply would not work. Too many vignerons, they argued, would refuse to join. The president of the Comice agricole, Paul Chandon de

Briailles, stated that even if many joined, there would be neighboring vignerons who would not participate, making the efforts futile. Vimont responded by noting that peasant sociability would be a means of disseminating information even to those who did not adhere. Conversations and neighborly gatherings would be effective. The high density of the villages and their semi-urban, semi-rural character seemed to support this position.[16] Others, however, simply scoffed at the idea.[17] There was little faith in the vignerons' ability to comprehend or address the problem. In the end, the idea of autonomous, voluntary associations was rejected. The Comice agricole, however, unanimously agreed that some type of organization was needed. What kind of organization and how it would be organized was left for future discussions.

While debate over how to proceed continued, Champagne's special "immunity" stemming from the terroir seemed to be confirmed by the onslaught of the pest in neighboring vine regions while the Marne remained untouched.[18] The public could follow the slow progress of phylloxera in the north and the miraculous immunity of Champagne. Professional journals, regional newspapers, and the local industry paper, the *Vigneron champenois*, regularly reported on new sightings of infected vines. While this information was disseminated in the media, the departmental Vigilance Committee did little to inform the rural population. Details came from the Vigilance Committee via the prefect of the Marne in the form of memos to mayors. Once these arrived in the villages, the instructions were simply for the update to be made available to interested vignerons.[19] Committee findings did not appear in local papers, and there was no attempt to disperse treatment information. The most "scientific" information (that is, coming directly from research labs or conferences) on treatment of the vine disease arrived from other sources: the national *Journal d'agriculture pratique* published a series of articles about the pest beginning in July 11, 1867; the first in a series of regional articles appeared in the *Vigneron champenois* on March 7, 1877, over a decade before the pest's appearance in the Marne; and in 1880, Vimont published his *Petit manuel & calendrier phylloxériques à l'usage des vignerons de Champagne*, a handbook providing up-to-date information on phylloxera in language that was accessible to vignerons.[20] These articles and the *Petit manuel* expressed the diversity and uncertainty of opinions on both the pest and treatment methods found throughout France.

Once phylloxera made its appearance in 1890 near Tréloup (Aisne), a few hundred yards west of the border with the department of the Marne, recriminations began. A number of observers blamed peasant ignorance or incredulity for the absence of strategies for fighting the pest. In *Six semaines en pays phylloxérés* (1896), a book on how to deal with phylloxera that was often cited, the local savant and Catholic leader Abbé Dervin, for example, argued that vignerons doubted that the Champagne vineyards would be attacked and had thus failed to study or implement any program of treatment.[21] It is difficult to judge the accuracy of his statements. Given the growing anti-clerical sentiments in the region in the 1880s, it is doubtful whether his rank in the Catholic hierarchy placed him in an ideal position to be familiar with the sentiments of the vignerons. He was correct in saying, however, that there was a certain passivity on the part of the rural community. But this passivity was not confined to any one segment of the population. The work of the departmental Vigilance Committee or the absence thereof suggests that skepticism and inaction were the norm. With almost a decade between the formation of the Vigilance Committee and the discovery of infected vines in the Marne, the committee and departmental administration, perhaps mirroring the factional disputes dividing the national ministries and reducing their effectiveness in the same period, missed a crucial opportunity to educate and mobilize the entire community.

Vignerons reported ailing vines throughout the valley of the Marne by the end of 1890.[22] Writing about attitudes in the region, one observer noted in the *Vigneron champenois* that "in Champagne we hardly believe in phylloxera."[23] With disbelief still widespread among all sectors of the population and no plan outlined for combating the pest, there was, not surprisingly, no community-based decision regarding treatment. Taking the initiative, Gaston Chandon, one of the largest landowners in the Marne and a member of the family firm of Moët & Chandon, purchased the first infected vineyards and destroyed the vines. Hoping to stop the destructive path of the pest, he adopted a system, first used in Switzerland, that involved inundating the infected vineyard with massive quantities of chemicals and then uprooting the vines. The "Swiss method," as it was known, effectively sterilized the soil, killing everything near the small patch of infected vines. There was then a short waiting period to see if the vines would send out new shoots. If no life reappeared, the plants were uprooted and the land remained fallow for several years before replanting began. This method was effective in areas where vineyards were small and relatively isolated. Even after this expensive treatment and subsequent replanting, however, the new vineyard was always susceptible to reinfestation.

Chandon's swift action appeared effective; there were no new discoveries of infected vines for several months. Subsequently, the Chandon family took the lead in the struggle against phylloxera, publishing some of the most detailed studies of the progress of the disease during the critical years from 1890 to 1894.[24] Having succeeded with chemical treatments, the Chandons became chief regional opponents of grafting with American rootstock, persuading the Syndicat du commerce to join them in promoting a rigorous campaign of chemicals and uprooting.[25] Chemicals, they believed, would stop the disease and make it unnecessary to replant with vines grafted to American rootstock, deemed unsuitable to fine wine production.[26]

By August of 1890, several members of the Syndicat du commerce approached the prefect, P. Viguié, with a proposal to create a *syndicat contre le phylloxéra* (anti-phylloxera association). The departmental council (*Conseil général*) and prefect were persuaded of the gravity of the situation. An administrative study of the feasibility of a departmental organization began under the combined leadership of the Syndicat du commerce and the departmental professor of agriculture, Edmond Doutté, in August 1890.[27] Several days after the decision to begin the study, the Syndicat du commerce sent a letter to the prefect proposing that any new organization dedicate itself to overseeing future chemical treatments.[28] This went against the general trend among growers throughout France. In the wake of the debates at the Bordeaux Conference of 1881 between proponents of chemical treatments and those who favored replanting with resistant American rootstock, growers in the south were opting to replant. Vimont, who was now a member of several national and international organizations for the study of the phylloxera, was also increasingly skeptical of the effectiveness of chemical treatments and the long-term efficacy of the Swiss method. As early as 1880, he had suggested that the use of resistant rootstock would be the only effectual method of dealing with the blight once it became fully established.[29]

Various obstacles initially hindered the use of American rootstock, however. The diversity of French vineyards required experimentation and field tests in each region to find American roots that could be adapted to the microclimate, soil conditions, and cultivation methods. Not all American vines were equally resistant to vine pests, and finding rootstock adaptable to very alkaline soil, like that found in the north around Champagne, was difficult. Until the invention in 1892 of the calcimeter, which measured the chalk content of soil, finding the best rootstock for the local soil took a great deal of experimentation.[30] Experimentation did not, however, appear to be an option in Champagne in 1890. Going against trends throughout France, the Syndicat du commerce banked on the success of chemicals. The biggest obstacle to experimentation with American plants in Champagne did not appear to be lack of time or resources but the négociants themselves. With the support of local officials, the négociants and Syndicat du commerce were taking control of treatment of the vineyards.

The sightings of infected vines in Champagne could not have come at a more inopportune moment for négociants. After years of petitioning the Ministry of Commerce and the Ministry of Foreign Affairs for the creation of an international agreement for the protection of regional brand names and the appellation "champagne" as uniquely French industrial products, a meeting of representatives of various nations to negotiate an international trade agreement was slated for the following spring, in 1891. This was the long-awaited Madrid Conference, organized around the signatories of the Convention of Paris of 1883. The Convention of Paris and the subsequent conference for revision held in Rome in 1886 had the adherence of all the major Western powers for the protection of commercial names (brand names such as Veuve Clicquot, Mumm, and Mercier) from counterfeiting.

There was no agreement, however, for the recognition of wine denominations—bordeaux, cognac, champagne—as protected industrial property. For the makers of France's luxury wines, such names were just as worthy of protection as the brand names themselves. France's internal laws were without effect in the crucial overseas wine markets. The delegates at the 1883 convention had taken up the question of appellations. But, as was the case in the French legal system, it was difficult to determine which denominations had become generic and which had retained a unique character. Indeed, the delegates abandoned the discussion when the debate turned to protecting everything from "velours d'Utrecht" to "eau de Cologne." Major exporters like the grandes marques firms of Champagne continued to lobby ceaselessly for the extension of the 1883 convention to wine appellations. There was particular pressure placed on ensuring that the Americans, large consumers of sparkling wine, become signatories to these conventions. With the appearance of the phylloxera, talk of replanting with American rootstock, and growing uncertainty in the wine markets, all of these efforts seemed jeopardized.

The swift response of the Syndicat du commerce in the Champagne vineyards and later publications to reassure consumers of the continued quality of champagne suggest that négociants feared a serious phylloxera outbreak, not only because it could jeopardize quality production and grape supplies, but also because it was a potential threat to industry claims to be manufacturing a unique, luxury product worthy of protection under domestic and international law. Several times in the 1890s, for example, the secretary of the Syndicat du commerce issued a small brochure to assure English-speaking consumers that phylloxera was not a threat to quality or to supplies of champagne. The booklet was concerned, above all, with linking "champagne" with the merchant-manufacturers and their brand names in the "old province of Champagne."[31] Producers, particularly in the Syndicat du commerce, had long promoted "champagne" as a superior product made from unique wines, which naturally commanded a higher price. Having stressed the connection between regional producers and quality in costly promotional techniques, négociants were eager to preserve not only the reputation of their product abroad but also the mystique of the geographic name.

The négociants of Champagne had pursued legal action against producers who allegedly counterfeited or imitated regional brand names and used the appellation "champagne" as a generic label, but their recourse was limited. Manufacturers in France and abroad, attempting to capitalize on the popularity of sparkling wine, imitated established brands or simply counterfeited their logo and used "champagne" as a generic label to describe the production technique that created the bubbles. The British writer Henry Vizetelly wrote in his work on the region, for example, that "the sparkling wines of the Loire are sold on the British market and elsewhere labeled Crème de Bouzy, Sillery, Aÿ mousseux, while their corks are marked with the names of fantom [sic] firms supposedly established in Reims and Épernay."[32] Benefiting from the reputation of known champagne producers, these wines were sold at low prices to less-affluent consumers. Regional négociants alleged that the "sparkle" was the only thing that resembled the original beverage, and sales of these beverages under the name "champagne" destroyed the reputation of established firms.[33]

Within France, merchant-manufacturers had considerable success in taking legal action against producers in Saumur who used the name "champagne."[34] French courts ruled in a series of decisions over two decades that "champagne" did not fall into the public domain as a generic processing technique, as in the case of "savon de Marseille" or "eau de Cologne." In the most important case, decided by the court of appeals in Angers in 1887, the court ruled that "champagne" had traditionally designated a natural product of the region that was rendered "sparkling" (*mousseux*) in a second processing phase undertaken in the region. "Champagne," the court determined, was not the generic name of this second processing technique.[35] This opened a new question: what created the natural product? With a highly manipulated, blended wine like champagne, it was ambiguous. Was it the varietal grapes? Was it a unique terroir? And, perhaps most important for the vignerons, if it was indeed terroir, what were its boundaries?

These questions were closely related to those being asked of other wine regions, particularly in the south. In the late 1880s, the "falsification" of wines—either with additives or by deliberate misattribution of their origin—had become a pressing national issue. New public health concerns grew in tandem with industrial food production and capitalist agriculture. Advocates of laissez-faire placed their faith in the power of free markets, the consumers' ability to make informed choices, and the ability of the judiciary to intervene when consumers faced dishonest producers or falsified products. Within the wine industry itself, there was a widespread demand for state intervention, a movement toward *dirigisme*. Faced with the potential damage to their reputations by falsifications, large winemakers, including those in Champagne, wanted state intervention to protect their products. Small wine producers, especially in the Midi, who were unable to keep pace with rising demand because of the falloff in production caused by phylloxera, also sought state intervention to halt the flood of falsified wines made by fellow vignerons, which kept prices low in the face of rising costs.[36] Public health advocates likewise put considerable pressure on the national government to intervene to protect consumers. The government of France reacted by creating commissions to study the problems related to falsifications throughout the 1880s. The Loi Griffe, which regulated the production of "falsified," sugared wines, was in place. The questions that concerned the luxury wine industries, particularly issues of "authenticity," had not, however, been addressed by early 1890.

These issues were particularly important for the champagne négociants when they arose outside France. The industry had no legal basis for preventing foreign producers from using the name "champagne."[37] According to the *Vigneron champenois* in 1898, Russia alone produced over a million bottles of sparkling wine labeled "champagne" each year.[38] But the *Revue des deux mondes* declared in 1890 that in imitations, "Germany is the leader, at least concerning the exterior of the bottles, with ornate French labels, invoking the villages and illustrative personalities of [Champagne's] viticultural history."[39] According to German law, however, any producer in Germany making sparkling wine that included wines from the Aube or the Marne totaling at least 51 percent of the finished product could legally use "champagne" on the label. Regional sources suggest that German buyers made yearly purchases in some of the best vineyards of Champagne.[40] By 1880, with producers of German sparkling wine, or *Sekt*, shipping nearly five million bottles a year, regional négociants in the Marne were anxious for international recognition of "champagne" as an *appellation d'origine* denoting a unique product of Champagne merchant-manufacturers.[41] Phylloxera threatened these efforts by raising more specific questions about grape origins and the role of terroir, all of which could hurt the reputation of the manufacturers. For the négociants, "treatment through extinction" seemed the only viable way to preserve the vineyards and regional manufacturers' claims to international protection of the "authentic" champagne.[42]

Government policy set by the Law of June 15, 1878, left most of the initiative for anti-phylloxera organizations to the local wine community. They were funded by local dues, which were then matched by the national government. The running of the syndicat and implementation of measures to combat phylloxera were left to local committees. To create an obligatory regional organization to coordinate and subsidize treatment, however, the law required the agreement of three-quarters of the local landholders. Mayors, deputy mayors, local *gardes-champêtres* (rural police), representatives of the Syndicat du commerce, and Doutté undertook a vigorous campaign to persuade the rural population to join the proposed anti-phylloxera syndicate in the Marne.[43] Doutté alone conducted forty-five local meetings between November 1890 and March 1891. Lasting from one to one and a half hours, these meetings drew between 80 and 160 participants. In Damery, home of the fiery René Lamarre, author of *La Révolution champenoise*, one meeting drew a crowd of over 350 vignerons.

Despite these intense efforts, the advocates of an anti-phylloxera syndicat encountered widespread resistance from vignerons.[44] According to one study of anti-phylloxera associations across France, resistance to organized treatment came mainly from vignerons with more than three hectares of land, who preferred to finance and implement their own approach to the pest. Those who did join syndicats came predominantly from the ranks of small owners (with an average holding of less than three hectares) who needed the public subsidies for treatment.[45] Vignerons in Champagne defied this trend. Most regional vignerons owned less than three hectares of land, but they only reluctantly joined the departmental organization. Most important, perhaps, those who were proposing and promoting the syndicat in the Marne came from the ranks of the négociants, including some of the area's largest landowners. Certainly, these wealthy property owners did not need the subsidies provided by the central government. Champagne thus followed a different model of opposition.

None of the original administrative circulars outlining the proposed syndicat have survived, so there is no clear indication of how it was presented to the rural population.[46] Evidence from petitions written by peasants and addressed to the administration suggests that there was a great deal of ambiguity surrounding the composition and funding of the organization and the proposed treatment methods. Vignerons would later argue that they did not believe that they were signing petitions to join any organization; they signed the petitions only after repeated assurances that "these signatures do not enlist members; we only want to know if there will be a majority in favor of establishing a syndicat."[47] In an administrative memo circulated to local mayors on September 27, 1890, however, Prefect Viguié said that a study done in August 1890 had indicated that the rural population was favorably inclined to join. Assuring the mayors that all vine owners would have the opportunity to participate at a later date in formulating the organization's constitution, he asked them to register names of all those joining in the *projet de Syndicat*.[48] It was perhaps this emphasis on registering for a *projet*, or plan, that created the ambiguity. Without a constitution or clear set of proposals, there was a sense, at least among the vignerons, that this was more a survey for a *projet d'avenir*, or future plan, than a membership drive for a concrete, imminent organization.

Apparently, the misgivings regarding the proposed association were fueled as much by the ambiguity in the petitions circulated locally as by the involvement of the négociants and Syndicat du commerce in its foundation. "The vignerons are above all hostile to the [syndicat] project, because they are suspicious of an initiative coming from the négociants [who are] considered responsible for their misery," a representative of the Syndicat du commerce wrote to the prefect.[49] The *Vigneron champenois* echoed this belief, saying that tensions and suspicions could create a "war" between the small growers and the négociants.[50] In a report on the general hostility toward the proposed syndicat, the subprefect noted that in the town of Vincelles, a majority of vignerons were opposed to it because they believed that the Syndicat du commerce had paid too little for the last harvests.[51] The mayor of the village of Dormans reported to the prefect that only eight local vignerons would sign the petitions, because they believed the organization would privilege those who controlled the *grandes crus*—that is, primarily the négociants.[52]

Vague proposals for what appeared to be a négociant-sponsored organization came amid economic hardship and growing social tensions within the champagne industry. Vignerons at organizational meetings for the anti-phylloxera association throughout the region expressed hostility toward the proposed syndicat.[53] René Lamarre had expressed similar sentiments in his pamphlet *La Révolution champenoise*, issued only two months earlier. "We are [the négociants'] chickens with the golden eggs, and they are killing us!" Lamarre wrote. "And our vines, our poor vines; you know very well, Monsieur, that if phylloxera is not in Champagne, it is thanks to the care that we provide.... This care we cannot provide this year, and phylloxera is at our door!"[54] Echoing those experts who believed that aphids attacked only already weakened vines, Lamarre argued that vignerons could not provide the necessary care to continue warding off the aphid because of the economic hardship caused by plummeting grape prices in 1890. Not unlike the peasant in figure 16, Lamarre identified the devastating parasites sucking the life from the vineyard community as none other than the bourgeois négociants.

Vigneron opposition to the anti-phylloxera initiative coalesced into a loose-knit protest movement. Although there was no central organization or leadership, Vimont, whose earlier organizational efforts had been rejected by the administration, and Lamarre, whose cooperative initiative had recently been launched, emerged as vocal spokesmen for the vignerons' grievances. Vimont had originally embraced chemical treatment. But, as he had clearly stated almost five years earlier, there was little reason to continue with chemicals once the phylloxera had become established. Indeed, he was one of the few who believed Champagne to be uniquely *vulnerable* to the pest. "Our methods of cultivation, *far from working against the spread of phylloxera, as is so often repeated, will make [the vineyard] more vulnerable*," he wrote in his *Petit manuel* (emphasis in the original).[55] Vimont, who called himself one of the "most sincere and devoted" friends of the vignerons, continued to promote a combination of individual effort and voluntary, autonomous organizations.

Lamarre, through both his activism in villages and his conversion of *La Révolution champenoise* into a weekly newspaper, emerged as a more radical voice in the vineyards.[56] He continued to advance his cooperative as the long-term solution for vignerons, but he also focused attention on the immediate threat of phylloxera. The *Révolution champenoise* published articles, some by noted wine specialists, on the efficacy of various treatment methods.[57] But most of his efforts were directed against the proposed syndicat, which, he argued, would serve as a tool for increased négociant control of the terroir of Champagne. Decent grape prices would allow vignerons to care for their vines and minimize the spread of phylloxera. The chemical treatment methods used by Chandon, he said, were the equivalent of "calling the doctor when the patient is dead."[58] For Lamarre, the solution was clear: a dual struggle had to be waged against the phylloxera and the Syndicat du commerce. Much like Vimont, Lamarre argued that the rural population needed to unite to form their own loosely centralized, locally controlled organizations.

While debate raged over the proposed syndicat, more infected vines were discovered. Amid tension and confusion regarding property records and the signature collection procedures,[59] Prefect Viguié declared that he had a majority of signatures for the new organization. The framework of the proposed organization was being studied, he informed the minister of agriculture; the vignerons had agreed to the proposal, and "17,500 proprietors out of 26,000 have asked to take part in the association." With this, he called the first general meeting of the new organization—a decision that would divide the community at a critical moment.[60]

The first assembly of *propriétaires-vignerons* for the Association syndicale convened in Épernay on the afternoon of July 11, 1891. According to Lamarre, the choice of meeting time—midday during a critical period in the growing season—and the lack of publicity for the event resulted in low attendance by vignerons.[61] The prefect reported to the minister of agriculture that 1,200 vignerons had attended the meeting; leaders of the vignerons disputed this number, arguing that only 300 vignerons had been present.[62]

Controversy surrounded the Association syndicale from the start. The prefect's failure to release the minutes of the meeting upon request led to accusations that the administration and the négociants were secretly conspiring.[63] The minutes reveal clear vigneron opposition. One vigneron who attended the meeting, M. Aubert from Aÿ, immediately called for tabling any vote on the statutes for the syndicat until more vignerons were present. But Prefect Viguié apparently saw no reason for delay and pushed the meeting forward.[64] The first item on the agenda was the constitution and regulations. Statutes for the new organization were promulgated swiftly and with little debate. This was perhaps because these statutes appeared remarkably democratic. All regional vignerons had the right to vote regardless of their total landholdings. Representatives of absentee proprietors could also vote if those representatives had a minimum of one hectare. Participation was granted to all, whether they had signed the original membership forms or not.

Further examination of the statutes, however, reveals that the syndicat was not to be strictly democratic. Those with more than one hectare, a small but powerful elite, which included a large number of négociants, were given additional votes. They were to receive one vote for each additional hectare they owned, not to exceed a total of five votes. Raoul Chandon, a large landowner from Épernay, proposed an amendment to provide ten additional votes for each 100 hectares held by any one proprietor. This would have given the Chandon family alone nearly 100 votes. Although Chandon's amendment was rejected, the statutes requiring tenant growers to have one hectare (larger than the average holding in Champagne) and giving large landholders additional votes still favored the large landowners' control of the anti-phylloxera syndicate.[65] Large landowners thus had the means to become the arbiters of phylloxera policy for the entire wine community.

Financing the syndicate was equally controversial. A system of dues was adopted based on the average value of each vineyard: 3 francs per year for vineyards valued at less than 4,000 francs per hectare, 5 francs per year for vineyards valued at 5,000 to 6,000 francs per hectare, and 10 francs per year for vineyards valued at 10,000 francs per hectare and above. With grape prices consistently low, some vignerons argued, however, that payment should be proportional to the value of the annual harvest, not land values. Others wanted the elevated costs of maintaining cultivation taken into account. Large landowners, it was argued, often owned vineyards valued at 50,000 francs per hectare that earned 1,000 francs or more *la* pièce but were required to pay the same 10 francs in dues as the small owners. Still others maintained that charges should fall more heavily on the larger landholders, whose extensive landholdings, scattered among many communes, put them more at risk. One amendment challenging the payment schedule was presented to increase payments for landholders with land values of above 10,000 francs per hectare, but the amendment was rejected on the grounds that the current system was fair and equitable. In the eyes of the vignerons, the burden for funding négociant-initiated and controlled treatment methods would fall disproportionately on the small growers.[66]

The voting structure and financing of the new organization gave Lamarre powerful ammunition for renewing his attacks against the négociants. Focusing his attention on the German origins of some of the most famous regional champagne manufacturers—such as Kunkelmann, Mumm, Deutz, and Geldermann—he argued that the négociants were foreigners usurping the patrimony of the champagne community and, by extension, the French nation. Referring to the négociants as *Choucroutemann* [sic] (i.e., "Krauts," literally, "Sauerkraut men"), he argued that they were little more than foreign feudal lords who wanted to turn the vignerons into serfs by buying up the terroir and monopolizing champagne production.[67] With his thinly veiled references to *la revanche*, Lamarre incited vignerons to fight against the "coalition of millionaires, almost all of them from Germany."[68]

Lamarre's anti-German invective, isolating the négociants as outsiders, had a powerful appeal in the Marne, where the theme of "la Champagne aux Champenois" was standard political rhetoric. One of the most important qualities for electoral success in the Marne between 1871 and 1914 was for a candidate to be "Champenois." When the nation was challenged by internal divisions during the Dreyfus Affair, for example, rural voters responded to candidates who claimed to "share" the life of the rural electorate through connections to the land and through military service, proving themselves to be both favorite sons of the region and a patriots.[69] In a region where fidelity to the republic was a major electoral theme, it was easy to attach the label of "foreigner" to the négociants, with their German names, aristocratic titles, and political affiliations to antirepublican candidates.

Vignerons expressed their discontent, sometimes echoing the words of Lamarre in the pages of both the *Révolution champenoise* and *Vigneron champenois*.⁷⁰ Police records indicate that the authorities were deeply concerned about the agitation and growing unrest in local villages.⁷¹ The first successful formation of a vigneron cooperative to collectively represent the producers in negotiations with the négociants over grape prices—the Association syndicale des propriétaires-vignerons de Damery—fueled the vigneron protest movement.⁷² By the time Prefect Viguié called for a second general assembly in August 1891 to elect a steering committee, there was a loose-knit peasant protest movement determined to elect vignerons as leaders of the syndicat. Through Lamarre's journal, it promoted a list of peasant candidates against the "official" list of the prefect and the local administration. If they could not stop the formation of the Association syndicale, Lamarre and his followers were determined that it should be controlled by the vignerons.

The elections to the steering committee of the Association syndicale were an overwhelming success for the vignerons. The most popular protest candidate turned out to be the "devoted friend" Vimont, who obtained 2,625 votes, while the nearest "official" candidate, the négociant Gaston Chandon, received less than half that number. The "official" list included notables (some of whom possessed no land in Champagne *viticole*) and négociants, while excluding candidates from among the ranks of propriétaires-vignerons.⁷³ Ignoring the growing social tensions, the official candidates had used local journals to make paternalistic appeals for the support of the wine-growing community.⁷⁴

The list of protest candidates presented by Lamarre and the vignerons in the *Révolution champenoise* was dramatically different. Nearly half of the names were hyphenated compounds, such as Petit-Fresnet, Mignon-Oudart, and Orban-Savart, common among peasants of the region.⁷⁵ Extensive family networks and rural custom made it common practice among the vignerons to combine the family names of spouses.⁷⁶ Names of "official" candidates might be widely recognized among the rural population as "notables," but the compound names created another kind of recognition: identifying the protest candidates as small growers from within the region. The contrast between the protest list and the official list created the impression that official candidates were outsiders, not acting in the interests of the community, and foreigners to the vineyards of Champagne.

With the defeat of the official list, the protest candidates, led by Lamarre and nearly 3,000 chanting demonstrators, arrived at the next meeting of the Association syndicale prepared to take control of the steering committee. When the prefect delayed turning over control of the meeting, protestors stormed the podium. Within minutes, police poured into the building, dispersing the crowd. No arrests were made, but the meeting was officially "postponed."[77] Fearing more confrontations, the prefect refused to reconvene the syndicat. It was not until November 24, 1891, three months after the elections, that the protest candidates succeeded in meeting with the prefect, but it would take an additional two months and the intervention of the minister of agriculture before the steering committee was reconvened.[78]

During the months that followed, accusations were hurled at négociants. With phylloxera spreading and grape production falling, vignerons were stunned to learn that sales of champagne continued to grow. In the ten years that followed the appearance of phylloxera in the Marne, champagne sales abroad dipped slightly, from 21.6 million bottles to 20.6 million; by contrast, domestic sales jumped from 4 million to 7.4 million bottles. Questions were raised about the origins of these wines.[79] A much publicized and extremely damaging report by the Laboratoire municipal of Paris published two years earlier took on amplified significance. Considerable amounts of falsified "champagnes" had allegedly been sent to Paris from the Marne in 1889; the Laboratoire municipal promptly publicized the findings in order to protect consumers and public health. The Syndicat du commerce responded with a flurry of letters to the minister of commerce protesting this report and demanding a retraction from M. Girard, the director of the lab. The Laboratoire stood its ground, arguing that the wines were "egregiously" altered with wines from outside the region. National papers hinted at an attempted cover-up by the négociants.[80]

Angry vignerons used these findings, among others, to bolster their accusations. Exasperated négociants vehemently denied any fraud. It was in this tense atmosphere that the prefect reconvened the steering committee of the Association syndicale on January 30, 1892.[81] The twenty-five peasant representatives were shocked to find a reconstituted steering committee with an additional sixty members from organizations such as the Syndicat du commerce and the Reims Chamber of Commerce. The appointments of representatives from private organizations that donated to the Association syndicale were perfectly legal. The Conseil général, for example, gave the Association syndicale a grant of 40,000 francs. In return, the administration was required to give members of the Conseil général positions on the steering committee in proportion to its contribution.[82] The Reims Chamber of Commerce appointments alone ensured that the unelected candidates on the official list, as well as a number of négociants, had seats.[83] On the grounds that the expanded steering committee was now too large to meet regularly, it elected a smaller executive committee in March of 1892.

With only three elected vigneron representatives included in the executive committee, vigneron control of the Association syndicale was liquidated.[84] These maneuvers, however, rather than weakening the vigneron protest movement, spurred moderate vignerons to action. Rallying around Vimont and the protest candidates, this group appealed to the courts for redress against the Association syndicale. The vignerons' representatives argued that the committee's composition, the syndicat's methods of financing, the irregularities in the initial petition drive, and, in fact, the very existence of the Association syndicale violated the law.[85]

While the vignerons' representatives pursued legal action, the executive committee of the Association syndicale, under the direction of Florens Walbaum, the president of the Syndicat du commerce, proceeded with treatment of infected vines. Following the precedent set by Chandon, the committee voted to pursue chemical disinfecting and uprooting, maintaining the ban on the introduction of American rootstock into the region. Arguing that "the champagne vineyard is a national patrimony, and France has the right to protect it," the Association syndicale decided that preserving the established vines, at any cost, was its responsibility.[86]

Vignerons emphatically disagreed. Vimont and his faction of moderates took the lead in advocating experimentation with American vines.[87] They argued that the financial burden of chemical treatments was too heavy for small growers to carry and had in any case become outmoded.[88] Support for the position of the vignerons' representatives came from Doutté, the agriculture professor who had been instrumental in the creation of the syndicat. Although he had once been one of the most radical opponents of using American rootstock, a trip to the Midi and the Charentes region convinced him that replanting was the only long-term solution.[89] The Swiss method of eradication was a policy that had long been abandoned in other regions.[90] Frustration could only mount, he argued, along with other observers, because growers employing chemicals against phylloxera were like "children . . . who attempt to stop the sea . . . with a weak barrier of sand."[91]

Vimont pushed the situation to the brink in a passionate speech to the Association syndicale and hundreds of vigneron supporters in which he called for the immediate replanting of infected vineyards with American rootstock. When Alfred Werlé protested that the association had a duty to "protect" the Champagne vineyards, the vignerons' representatives shouted for the convening of a general assembly to decide the issue by popular vote. Amid cheering vignerons, Vimont roared that there was only one protector of the vineyards—the vignerons. The ensuing pandemonium was eventually quelled by the arrival of police.[92] Battle lines were clearly drawn.

Local officials expressed concern that the debate had "overexcited" the rural population.[93] Far from seeing the situation as "overexcited," however, vignerons and their representatives complained of a general inertia. Continued refusals on the part of Prefect Viguié and the Association syndicale leadership to call a general assembly of propriétaires-vignerons led to the resignation of vigneron representatives.[94] A "cold war" ensued in local villages. Vignerons shunned fellow vignerons and villagers who cooperated with the syndicat. According to one vigneron employed by the Association syndicale: "It is impossible for me to continue any longer to be the foreman, because everyone is angry with me, to the point that I do not dare to show myself [in public], and if I continue working [for the syndicat] I believe that I . . . would be obliged to leave Vincelles; in short, I am regarded as a convicted criminal."[95] Militants like Lamarre called on the rural population to resist the "flooding of the vines with poison" by "energetically employing all means" to stop the Association syndicale.[96]

Vigneron frustration culminated in violent confrontation on the eve of the harvest of 1892. In the town of Damery—home of Lamarre and the rival Association syndicale des propriétaires-vignerons de Damery—peasants were furious. Rumors circulated that négociants would boycott any vignerons taking part in Lamarre's new cooperative. Cultivating tools became makeshift weapons. Armed with pitchforks and *fusées paragrêles*, angry vignerons prevented an eradication team from the Association syndicale from entering the vineyards. Undeterred, the team returned that night and uprooted infected vines, destroying an adjacent family garden with doses of chemicals. By the early morning hours, news of the clandestine eradication had spread to neighboring vine-growing villages. Villagers hurled accusations that this was the work of "outsiders" who had willfully violated peasant "custom"—entering the vineyard at night and without the proprietor being present. Even though these "traditions" regulating behavior in the vineyards were new and did not predate the nineteenth century, they became part of a rich history in the context of struggle. Claims that the eradication team had violated custom had less to do with tradition than with the balance of forces in the struggle over the Association syndicale.

In the days that followed, irate and exasperated vignerons chased syndicat work teams from the vineyards of five other villages (Venteuil, Cumières, Fleury, Boursault, and Vauciennes), firing their hail rockets and assaulting officials with agricultural tools. Treatment teams were barred from entering vineyards throughout the valley of the Marne; vignerons refused to pay their dues to the Association syndicale, citing its illegality and lack of vigneron representation.[97] Benefits were organized by Lamarre to raise money for the vigneron family whose vines had been destroyed in Damery.[98] Even the most moderate vignerons could not ignore the blatant disregard for rural tradition.[99] Referring to "vandals" and "savages," Lamarre highlighted the destruction of "our vegetables, which were our only resource."[100] In destroying the meager harvest of the vines along with the adjacent family garden, the Association syndicale appeared indifferent to the very survival of the peasants.[101]

Négociants and members of the Association syndicale were not silent during these attacks. Peasant ignorance was the true threat to the community, they argued, and the real source of vigneron resistance. In an interview that appeared in the Radical newspaper *L'Indépendant rémois*, Alfred Werlé, a member of both the Syndicat du commerce and the steering committee of the Association syndicale, stated that violent clashes proved that the vignerons were against "science and reason" and motivated by "ignorance and prejudice." He expressed amazement that the vignerons did not "understand their true interest," which was to follow the lead of the association. If the vignerons relied on their own resources, he argued, they would ultimately be ruined.[102] Werlé's statement implied that the large landowners had little need for the association to treat their own vines; they were, rather, "moved by the best intentions."[103] Those best intentions were echoed by others who believed that the administration needed to use more coercive methods to curb vignerons' defiance.[104] The editor of the *Vigneron champenois* joined négociants in denouncing opposition as nothing more than peasant stupidity.[105] Charges of peasant apathy and skepticism legitimized the decision to dismiss vignerons' objections: only the "enlightened" leaders could protect the interests of Champagne.[106]

Petitions presented to the prefect in 1892 regarding payments to the anti-phylloxera association contain numerous handwritten observations by vignerons demonstrating that some, in fact, did not believe that the blight existed. In the commune of Mareuil, for example, it was asserted that "l'existence du philoxéra [sic] n'est pas suffisamment prouvée," and even that the there was, in fact, no such thing: "le philoxéra [sic] n'existe pas, c'est une maladie imaginaire."[107] Evidence from the same town, however, indicates that another, much larger segment of the community was relatively well informed about both the dangers of phylloxera and the advantages of various treatment techniques, expressing opinions on these methods and the distribution of costs for the Association syndicale.[108]

Doutté, who worked closely with the rural community, argued that indifference arose less from ignorance than from a misunderstanding of the gravity of the situation.[109] As phylloxera spread and vignerons' alarm increased, the vignerons' voices, recorded in petitions to the prefect, reveal a profound concern and desire for action.[110] The action desired, however, ran counter to the "chemist" position of the négociants. Peasants wanted to replant. In one petition presented to the mayor of Épernay, a group of vignerons wrote: "Phylloxera is at our doors and a large part of the vines are anemic and sterile. We need to regenerate our vines . . . and to profit from the experience of others before we lose everything."[111] According to Vimont, the vignerons carefully monitored the spread of phylloxera in the region and simply saw no reason to adopt chemical treatment when it had failed in other areas of vine cultivation. The vignerons, he wrote, "were therefore neither ignorant nor did they misjudge the situation."[112] Moreover, evidence suggests that, just as Vimont had predicted, articles published in the *Révolution champenoise*, covering the benefits of various treatment methods, were debated and discussed in the vineyards.[113] Ignorance and indifference were convenient excuses for dismissing peasant opposition.

Over the next year and a half after the 1892 resistance, peasant opposition continued, with small-scale confrontations between vignerons and representatives of the Association syndicale.[114] Local police were increasingly concerned that vigneron discontent could become violent and more generalized, particularly in the region around Épernay.[115] By the beginning of 1894, the slogan "The vine for the vignerons!" was found scrawled across walls in villages throughout the region.[116] Police estimated that nearly 75 percent of vignerons were actively opposed to the association.[117] By February 1894, peasant resistance had once again escalated into armed confrontation.

Entering the town of Vertus in the Côte d'Avize, a representative of the Association syndicale called on vigneron families to collect overdue association payments. Although Vertus was far from the areas of protest in the valley of the Marne, the representative nonetheless brought an escort of gendarmes. News of this spread quickly, inflaming peasant animosity. A hostile crowd interrupted the private proceedings in the courtyard of a peasant home with shouting and shaking fists. As rocks rained down on them from all directions, the representative and the gendarmes were forced to flee into the street. When a gendarme fell and was reportedly kicked by irate protestors, his companions drew their sabers and scurried to the shelter of the *hôtel de ville*.[118] From the safety of the mayor's office, they wired for police support.

The protestors dispersed, only to return moments later, carrying pitchforks, hoes, and hastily fashioned arms. They milled around the hôtel de ville, where they were joined by neighboring villagers. A tense standoff between police and protestors continued through the night. In the morning, protestors presented their demands: dispersal of the police and the resignation of the mayor, who they believed had collaborated with the syndicat conspiracy by coercing peasants to give their initial signatures to the project back in 1890. The mayor resigned; the police retreated, later claiming that they had had little choice after an unnamed informant reported the presence of a bomb and dynamite among the crowd. The jubilant crowd thereupon escorted the association representative out of the village.[119]

Events in Vertus brought fears of more violence in the largest vigneron villages, particularly in the neighboring communes of Avize, Cramant, and Oger.[120] The victory at Vertus appeared to mobilize the peasant community.[121] It was the moderate faction of the rural community, based mainly in the villages of the Côte d'Avize, that emerged to direct the protest movement in the wake of the Vertus confrontation. In the valley of the Marne, militants like Lamarre were distracted by work for the first producers' cooperative, established in Damery in 1893. The moderate faction, organized around Vimont, collected signatures for the dissolution of the Association syndicale, which they presented to the minister of agriculture in March 1894. Calling on Albert Viger, the Radical minister of agriculture, who was noted for his pro-peasant policies, they succeeded in forcing a confrontation with members of the executive committee of the Association syndicale and the departmental administration. Hours of negotiations ended in a compromise. Both sides agreed to the formation of locally controlled commissions for vine treatment, although the original Association syndicale would be left intact until the end of its tenure in 1895. Local commissions had effective control over treatment policy, but members of the administration and the steering committee of the Association syndicale were not humiliated by the total dismantling of the central organization. Amnesty was then extended to those arrested or fined during the protests in Damery and Vertus.[122]

Years of struggle against the Association syndicale thus ended in vigneron victory. But rather than holding jubilant celebrations, the vignerons were silent. Perhaps exhausted from the stormy encounters and heated rhetoric, the rural community turned its attention to the laborious task of uprooting vines and replanting. Although members of the central organization grumbled about commissions in local communes, more vines were uprooted or treated in 1894 than in any year during the association's existence.[123] Even the characteristically outspoken Lamarre was silent, called away by Marianne for his obligatory military service. Almost without notice, the Association syndicale was dismantled in 1895.[124]

"I will not waste my time nor my money defending myself against phylloxera," one vigneron stated unequivocally. He was not backward or ignorant; indeed, he was following the lead of vignerons throughout France who had struggled for years against the blight. As the pest spread in the late 1890s, this sentiment was echoed by many other small growers.[125] Michel Hau has noted that the phylloxera crisis was not "a passing accident, but a profound modification of the orientation of the agricultural production of Champagne."[126] The cost of phylloxera treatment and replanting, combined with the plummeting prices of the 1890s, sealed the fate of those few vignerons still producing ordinary red wines. In the commune of Broyes, for example, a small vine-growing area in the south of the department, vine growers engaged in producing *vin ordinaire* who had not capitulated to the pressures of the capitalist market now capitulated to phylloxera. Vines were systematically uprooted; vineyards were sold and, in some cases, converted to fine wine production.[127]

Given the enormous costs of replanting and the other economic difficulties confronting the vignerons, it seems highly unlikely that the majority of peasant growers had the money to make the investment in conversion to fine wine production or to purchase new plots of lands that became available. As one vigneron said years later, "And then they [the vignerons] went through the phylloxera invasion, which destroyed the vine in Champagne, as everywhere else. They were helpless to deal with mildew and other maladies of the vine . . . that life was difficult, very hard."[128] A report requested by the prefect of the Marne in 1907 on the "dividing up of the vineyards" suggests that there was a growing accumulation of property by the négociants in the wake of the phylloxera.[129] Vineyard prices were plummeting. In the area around Cumières, by 1900, vineyards that had sold for 30,000 francs per hectare before the phylloxera now sold for 6,000 francs per hectare.[130] By 1909, these vineyards would be worth little more than 2,200 francs per hectare.[131] Evidence suggests that these lands were purchased by négociants. By 1907, the sixty négociants and their families (who made up less than 1 percent of the vineyard population) controlled 22 percent of the vineyards surrounding Reims. Another 1,000 proprietors had holdings of between one and five hectares, amounting to about 32 percent of the total vineyard area. The bulk of the vignerons—7,340 proprietors, who made up over 85 percent of the rural population—possessed less than one hectare each, or about 50 percent of the total vineyard area. Some négociants, such as those in control of the maison Veuve Clicquot, did not substantially add to their holdings during the early years of the crisis in the 1890s, but by the early 1900s, they were making extensive purchases, often exchanging parcels to create large, unified vineyard holdings.[132] The discrepancy between vigneron and négociant control of champagne—both the wine and the region—was growing.[133]

The phylloxera crisis was in many ways a revolution for the wine-producing community of Champagne. Not only did it end the production of ordinary wine in the region and bring about a redistribution of property, but it also led to the profound modification of traditional cultivation techniques. The practice of planting *en foule*, for example, was not adapted to the use of grafted plants. Ill-suited for the traditional propagation method, grafted vines were instead planted *en ligne*, significantly altering the vignerons' work routine. Planting vines at regular intervals in long rows rationalized cultivation, allowing the introduction of animals and, eventually, machinery. Animals and machinery, in turn, eliminated some of the more backbreaking labor, such as hauling fertilizer. The neat rows of vines reduced soil erosion, abolishing the constant collection of eroded earth, and made the application of fertilizers and pesticides easier and less time-consuming. Ultimately, these changes boosted yields. By the early

1900s, vignerons who replanted *en ligne* with grafted vines using new cultivation methods reported an increase in average yields from twenty-eight hectoliters of wine per hectare during the pre-phylloxera period to over thirty-eight.[134] Grafted vines *en ligne* had a shorter life span than those *en foule*, but all of the labor-intensive tasks of rejuvenating vines each spring were eliminated, and grapes *en ligne* were better exposed to sunlight, allowing a more consistent maturation, higher yields, and better quality. From planting to harvest, techniques introduced in the phylloxera period simplified the work routine, reduced the need for additional hands, made it easier to observe work in the vineyards, and increased production.

Replanting, the treatment preferred by vignerons, accelerated the reorientation of Champagne *viticole* toward grape monoculture for the production of sparkling wine. For growers prepared to struggle against phylloxera, government subsidies were a necessity. Reconstituting the vineyards was an enormous, expensive, and difficult undertaking, hardly worth the effort for those interested in common wine production. Not only was it necessary to uproot and replant, but an entire parcel of land frequently had to be modified and the soil completely reworked in order to accommodate grafted vines and planting *en ligne*. Plants had to be purchased and transported; growers had to learn to graft vine shoots and properly care for new plants. For the small grower, the subsidies offered by the government through central syndicats, like the one that had just been dissolved, were a necessity.

The movement toward a new anti-phylloxera association that could funnel subsidies to the growers was started spontaneously by vignerons of the region. Embracing an idea originally proposed by Vimont and Lamarre, vignerons organized independent local syndicats in their villages to share information about replanting and phylloxera's progress.[135] By 1898, a small group of vignerons from these local syndicats and a number of négociants who had not participated in the Association syndicale took the joint initiative to organize a new anti-phylloxera treatment organization that would qualify for government subsidies. Almost three years after the demise of the Association syndicale, a loosely knit body, the Association viticole champenoise (hereafter AVC) was created for experimenting with American rootstock, purchasing chemicals and insecticides, and replanting.[136]

Why did the AVC succeed where the Association syndicale had failed? The new entity consisted of a general assembly, an administrating committee, and a technical commission jointly composed of vignerons and agricultural specialists. Controlled by vignerons, the anti-phylloxera association was no longer a focal point for conflict. Within the first year of its existence, nearly thirty locally controlled syndicats were grouped together under the umbrella organization of the AVC. Over the course of the next two decades, as phylloxera moved relentlessly through Champagne vineyards, thousands of vignerons would voluntarily join local branches of the AVC, participating in both its governance and its various educational programs.[137]

The creation of the AVC in 1898, it has been argued, facilitated better relations between vignerons and négociants following the bitter disputes of the early 1890s.[138] While the AVC appeared as a reconciliation of vignerons and négociants, only twenty-four négociants participated in the organization, contributing from 1,000 to 5,000 francs yearly. Some of the most active members of the Association syndicale—Chandon and Werlé, for example— did not belong to the AVC. Most glaring was the absence of the Syndicat du commerce from AVC membership. Publicly, the Syndicat du commerce accepted the AVC, stating that the interests of winegrowing and the négociants were identical. Writing that "the first duty of the proprietor is to struggle by himself to protect his property," the leaders of the Syndicat du commerce seemed resigned to the new decentralized association. Nonetheless, they continued to protest the use of American vines, refusing to support AVC efforts at replanting financially.[139] Ultimately, the Syndicat du commerce did not participate with the AVC until 1909, when it made a generous contribution of 389,816 francs to the AVC's coffers. This contribution did not, however, mark a period of Syndicat du commerce support for the AVC. Almost one-tenth of the 1909 contribution was ultimately dispersed by the négociants to proprietors outside the ranks of the AVC; the next contribution of the Syndicat du commerce was not sent to the AVC until 1913.[140]

Before the phylloxera crisis, the power of the négociants and the Syndicat du commerce to "speak" for the industry had never been challenged. As the manufacturers and marketers of champagne, they viewed themselves as the center of the champagne community, the arbiters and protectors of the regional economic and cultural legacy. From the statutes that favored the large landholders, many of whom were négociants, to the exclusion of vignerons from the steering committee, the Association syndicale reflected this négociant view. Control was meant to rest in the hands of négociants. Ultimately, it was the threat of escalating violence and increased government intervention that forced compromise. In the years ahead, however, it would become clear that stinging accusations of foreign origins and fraudulent production would moderate their collective approach. With no international body of law to protect the regional appellation and new challenges to their dominance on world markets, négociants reexamined their ties with rural France.

While the new AVC thrived and négociants withdrew from participation, other vigneron-based organizations of the 1890s languished. The producers' cooperative founded in Damery in 1893, the Association des vignerons champenois, Société coopérative de production, attempted to sell its first champagne production in January 1897 under the *marque* "Pur Champagne." Returning from his military service in 1897, René Lamarre renewed his leadership role, acting as the general representative of the cooperative. Faced with the same dilemma as firms that tried to break into the competitive market for luxury wines, Lamarre struggled to establish market networks and name recognition for "Pur Champagne." Using connections with socialists in the Parti ouvrier français, Lamarre reorganized the cooperative in 1898 and contracted to sell Damery champagne through Paris consumer cooperatives (*coopératives de consommation*). Despite the creation of a new label "Champagne des Trois-Huit," Lamarre found, champagne did not "oil the wheels of social life" among the working classes of Paris. Workers seemed uninterested in the cooperative's guarantees of quality and authenticity. The cooperative limped along for just over five years before being dismantled.[141]

When Lamarre returned from military service, he found that much of his militant support had dissipated. In the legislative elections of 1898, he ran as a self-proclaimed candidate of the vignerons against the Radical incumbent, Paul Vallé, a protégé of Léon Bourgeois and a popular supporter of protection of the champagne industry,[142] and Paul Coutant, a reactionary candidate accused of being allied with the Chandon family. The so-called candidate of the vignerons had little support outside Damery, garnering only 2,600 votes. The rural population voted overwhelmingly for Vallé, returning him to office for the third time. Even Coutant received more votes than Lamarre, mainly from within the city limits of Épernay.[143]

The impassioned rhetoric and vigneron militancy of the early part of the decade was gone. Yet the struggles of the phylloxera crisis and the years of contention resulted not in a retreat to fierce peasant individualism but in a renewed interest in the benefits and power of collective action within the Third Republic. Vimont and Lamarre had raised the idea of locally based peasant syndicats, and the success of the AVC proved the effectiveness of rural solidarity. Most important, however, peasant activists challenged the legitimacy of négociants' claims to control the economic and cultural legacy of champagne. The anti-German invectives and the effectiveness of calling on the state to back a vigneron-controlled vineyard association focused attention within France on the connections between the wine and the nation. Contention in the 1890s was the first in a series of efforts to define the boundaries of terroir (and with it the location of the "soul" of Champagne) and membership of the champagne community within the larger entity of France. The vignerons' willingness to join in collective action gave them a powerful weapon for future challenges to their place within the circle of production connected to the terroir of Champagne and their control of sparkling wine as a part of France's patrimony. Like the peasant in figure 16, the vignerons of Champagne showed themselves willing to take action, even violent action, against "the so-called phylloxera."

CHAPTER

BOUNDARIES

The Limits of the "True"
Champagne, 1900–1910

"The wine of Champagne, which must come from the former province of that name to qualify [for the denomination], includes within its blends only the wines of the Marne, and nothing more," wrote Raphaël Bonnedame, the editor of the *Vigneron champenois*, in 1899. Champagne was essentially a wine grown and transformed within the department of the Marne. Bonnedame, whose words were reproduced in pamphlets distributed to consumers throughout France, went on to explain that even among the vineyards of the Marne, there were only a few areas that had grapes used in making the "grands vins de Champagne." There were, of course, *crus secondaires* grown in the department that were used for other, lesser sparkling wines. Bonnedame was at pains to explain why these departmental boundaries, less than 100 years old, were at once unconditional and yet mutable. The Marne had clear geographic limits established after the Revolution. Yet vines stretched across these boundaries into the neighboring departments of the Aisne and the Aube. Grapes from these vineyards were incorporated, as Bonnedame reluctantly notes, into "certain wines of Champagne." But, he added, this was only in those years "where their exceptional qualities make it possible to use them for making sparkling wine." All other grapes from all other vineyards, outside of this ill-defined exception, within the prerevolutionary province of Champagne—those of the Aisne, Haute-Marne, Seine-et-Marne, and Aube—"despite their pretensions . . . cannot claim the honor of being a part of what we call Champagne *viticole*."

These extraordinary vineyards of the true Champagne *viticole*, Bonnedame continued, were famous because of "a fortunate combination of numerous advantageous qualities: their fine vines; their chalky soil; their unrivaled exposure [to the sun], all joined with the continuous special care that is lavished on them each day." Although Bonnedame did not use the term *terroir*, he was clearly working with the concept. Having addressed the combination of soil, sun, and cultivation methods, he made the critical link with "soul." "The wine of champagne has, without a doubt, transmitted its virtues to the Champenois," he explained. "The Champenois are relaxed and courteous—always an effect produced by champagne." This unique relation remained unchanged even with the devastation of the phylloxera, he continued, because of careful replanting by the virtuous Champenois with "choice French plants." The end result was a *vin gaulois* without rival for expressing the *esprit gaulois*.[1]

This Gallic wine could also be said to have expressed, as Bonnedame so carefully hints, a particularly troubled French esprit by 1900. The source of the problem was "fraudulent" production or defining the "true" champagne. From a rather peripheral concern in the 1890s, fraud emerged as the central issue for producers of both ordinary and fine wines at the turn of the century. Seemingly in unison, the French wine industry made demands on government agencies to deal with fraud. The state agencies, often divided internally and unsure about how to proceed, met these demands with increasingly complicated government legislation. Faced with the bellicose demands of the wine industry to make an ambiguous "other" party within the production process responsible for fraud, the government of the Third Republic seems to have been amenable but complacent, passing legislation that does not appear to have been seriously enforced.[2] As Charles Warner explains, the French government was "not likely to favor legislation adversely affecting the sale of anything labeled wine" during the Belle Époque, when it was spending millions to aid viticulture as it struggled to rebuild after the ravages of phylloxera.[3]

Most studies of fraud have focused almost exclusively on southern production and the power of the "wine lobby" dominated by the Midi growers. Very little attention has been devoted to fraud in the fine wine industry. Discussions of fraudulent production in fine wine are approached as an extension of the more global alteration problems in southern ordinary wine production. Given the early battles over the Griffe law of 1889, which concerned the use of sugar and water in wine, and the spectacular riots in the Midi in 1906 and 1907, scholars interested in the wine industry understandably have devoted minimal attention to Champagne, in particular, during the first decade of the twentieth century. Yet the problems for fine wine producers were not those of ordinary wine producers. Central concerns in the Midi, such as additions of water, sugar, or *vins des raisins secs* (wine made from raisins), were not relevant to the fine wine industry. Fraud, in an industry that openly manipulated the wine, adding sugar in the second fermentation process, was not necessarily a matter of physical alteration of a finished product.

While southern wine growers demanded laws restricting the adulteration of wines, fine wine producers lobbied for very specific legislation to safeguard, not the "purity" of the common beverage, but what they believed was the reputation of their product. Within Champagne, the fraud that required persistent government attention consisted of three related issues: counterfeiting of names of manufacturers or brands; using the denomination "champagne" as a generic label for sparkling wine; or labeling a wine "champagne" when the grapes had been harvested or the wine manufactured outside of the region. Fraudulent use of brand names and counterfeiting, fundamental concerns of the champagne négociants, were vigorously pursued by both the minister of commerce and the minister of foreign affairs. International laws regarding industrial property promulgated at the turn of the century strengthened the ability of the French government and the négociants to pursue this fraud. Files from various government ministries suggest that representatives of the Third Republic took this pursuit seriously. With luxury wines one of France's most profitable exports, the government had a financial stake in these issues.

The French judiciary had by 1900 also advanced the cause of protecting champagne as French property, ruling that "champagne" was a denomination, not a generic label. Domestically, regional manufacturers had successfully garnered state recognition of champagne as a French regional product, although the boundaries of that region were, as Bonnedame explained, ambiguous. This work was bolstered at the international level by the signing of the Convention of Madrid, which also recognized "champagne" as French. Coming in the midst of the phylloxera crisis, a vocal prohibition movement in Western nations, and a growing demand for consumer protection with respect to food products, the Madrid arrangement was not uncontroversial. Several key markets, in particular the United States, were not included under the terms of the treaty. Despite these shortcomings, the French government and négociants had a powerful legal tool for ensuring that the name "champagne" remained a French possession.

It would appear that champagne had both national and international recognition as a French national good. Yet the issue of its place within the nation was far from settled. By 1900, controversy revolved around more precisely defining what champagne's French identity meant. What *was* champagne? Was it a blend of certain types of grapes? Was it a blend of grapes from an exclusive region? What were the boundaries of that region? How was it delimited and who would demarcate it? Was there a relation between these boundaries and the manufacturing of the wine? And, most important, what was the basis for these limits and boundaries? Science? History? Economic necessity? Tradition? For a highly manipulated wine like champagne, without a long history, these questions evaded simple answers.

These were essentially questions connected to appellations and delimitations. Consumers, in France and abroad, today take for granted the strict controls of the *appellation d'origine contrôlée*, or AOC (the *indication géographique protégée*, or IGP, of the European Union), which define a wine's content and linkage with a delimited territory. The strict industry production standards, now part of French and European law, resulted from the early twentieth-century struggles of Champenois, both vignerons and négociants, to stop fraud by establishing the boundaries or limits of vine cultivation and the production of wine.[4] While the issues of appellation and delimitation would quickly become central to the wine industry as a whole, they first emerged in Europe as an internal issue for the sparkling wine–producing community of Champagne *viticole*.

It was the young Lamarre who articulated the issue for those within the region. "The fraud that we want to combat consists of this," Lamarre wrote in *La Révolution champenois*: "merchants who, undeserving of the name Champagne, sell wines foreign to our soil." Fraud, according to Lamarre, was carried out in several different ways: "Either they [négociants] fabricate wine in their firms or in those of their intermediaries, the agents, with grapes grown in the Midi; or they bring in wines from Lorraine, Burgundy, Anjou, or even Germany."[5] Lamarre's remarks were not unique. The idea that fraud was being committed was as old as the vineyards themselves. Consumers in the eighteenth century, for example, had originally rejected champagne because they believed the bubbles were a trick to mask the wine's bad taste.

"The connection between geography and quality had become a widely accepted belief, if not to say myth, by the nineteenth century, and is still the gospel of enophiles," Loubère observes.[6] By the end of the nineteenth century, the commitment to the belief led to accusations from different quarters that outside wines were being used by Marne producers. A letter to the Syndicat du commerce from the Chamber of Commerce of Maine-et-Loire, for example, alleged that the amount of champagne sold by négociants from the region was five or six times higher than the quantity of wine produced within the Marne.[7] Although condemned by those inside and outside of the region, the practice of making "champagne" within the Marne with grapes or wines from outside of the region was not illegal. Rulings by French courts for the protection of commercial property were directed against merchants outside the region, particularly in Saumur, who labeled wines that were not from the region as "champagne."[8] There was no similar decision against Marne merchants who engaged in the same practice—a fact that did not escape the notice of those in Saumur. In 1887, for example, in one of the Syndicat du commerce's cases against sparkling wine producers in Saumur, lawyers for the defense argued that "a rather large number of manufacturers in the Marne are the first to make wines indiscriminately from wines purchased outside the region."[9] The idea of Champagne négociants producing sparkling champagne with wines from elsewhere became so commonplace that even in a *Dictionnaire populaire de médecine*, it was stated that champagne was made with wines from Lorraine and Burgundy.[10]

Champagne producers could not remain indifferent to these accusations. With the reputation of champagne and brand names linked to its quality and purity, few négociants within the region would have admitted to fraudulent practice. In the face of public allegations, the Syndicat du commerce vehemently defended the integrity of the industry. When the *Revue de viticulture* published an article alleging that regional négociants produced "champagne" in German cellars, for example, the Syndicat du commerce blasted the author, pointing out that its statutes required that all wines bearing the name "champagne" be from grapes harvested (*récoltés*) in Champagne and all aspects of production (*manutention*) be completed in the region.[11] While it was true that the organization had expelled one member who did not adhere to these rules in the late 1880s, members failed to mention that there was no mechanism for enforcing Syndicat standards on regional négociants. Production did, indeed, continue even after the merchant was expelled. Having stressed in expensive promotional campaigns that only sparkling wine produced within the confines of the Marne should be considered "champagne," the Syndicat attempted to deflect any bad publicity directed at regional champagne manufacturers. Although the Syndicat was ready to denounce fraud, it was much less willing to admit that local négociants engaged in it.

For the vignerons, it was Lamarre who defined fraud in terms of the grape producers of the champagne community in the pages of *Révolution champenois* in the early 1890s. In Lamarre's formulation, champagne derived its quality, its prestige, not from the firms in Reims or Épernay that bottled the final product, but from the land and the Champenois vignerons who tilled it. Lamarre proposed that vignerons rally together to protect their "property" through legal recognition of a "Champagne délimitée." By creating publicly recognized boundaries for "champagne"—both the viticultural region and the community of producers—the vignerons, he argued, would have a powerful tool for punishing unscrupulous merchants who sacrificed the rural community for higher profit.[12] "It is time that messieurs les négociants learn that their interests are intimately tied to those of the vignerons."[13] For Lamarre, the appellation of "champagne" meant nothing if it was not unequivocally linked to the terroir so eloquently spelled out by Bonnedame.

By 1900, both vignerons and négociants in the Syndicat du commerce were increasingly alarmed by reports that champagne négociants, without distinction, were using the regional denomination for wines produced with grapes from outside the region. Phylloxera was taking its toll: vignerons were faced with rising costs, the task of replanting (fully under way by 1900),[14] vineyard consolidation, and rationalization of production, which led to excessive grape supplies and falling prices. Many on the land faced economic ruin. Yet the number of bottles of finished wine sold and négociants' profits continued to climb. The foreign markets remained relatively stable, with yearly sales of between 20 and 23 million bottles between 1900 and 1910; meanwhile, the domestic market saw a doubling of sales from 1900 to 1910, going from a modest 7 million bottles in 1900 to 15 million in 1910. From St. Petersburg to Los Angeles, there was growing suspicion that the consumer was being enticed by the popularity and prestige of the word "champagne" into buying ordinary wines at inflated prices.[15] The basis of the concern about these figures and reports was very different for the vignerons and négociants, but the end result was the same: a desire for intervention by the French government for more aggressive protection of "champagne."

The response of the government to these demands was, at first, feeble and contradictory. Study after study was produced and endless debates were conducted on how to deal with the *mévente*, or slump in wine sales. The protective legislation that followed in the early 1900s has led Charles Warner to call the wine industry the "spoiled child" of French agricultural policy. There is a certain irony in this, however, because while the state was supposedly giving in to the demands of the "spoiled child," it was simultaneously working to turn "peasants into Frenchmen," in Eugen Weber's words. Urban public institutions and capitalist markets at both the national and the international levels functioned to integrate the separate *pays* of rural France into the nation, a process of domination of the periphery by center that has been called akin to colonization. The discussion of fraud—and particularly the connected discussion of boundaries—suggests, however, that the winemakers were neither spoiled children getting all that they unreasonably demanded from the French national family nor a conquered colonial people brought reluctantly into the modern world.[16] Indeed, the divisions within the French government, within the French wine industry as a whole, and among those in Champagne *viticole* proper hint at a more complex series of negotiations over the place of wine—as both a capitalist industry and a marker of Frenchness—within a larger discourse about France and French national economic interests. As the contention over fraud highlights, private companies and local peasant organizations promoted their regional specialty as a national good, integral to the common imagined past that was actively promoted by the state. Within the negotiations over appellations and delimitations was a discussion of the place of the regional wine community within the nation. The process of negotiating the limits of the "true" Champagne created a common culture in which local traditions became part of a national narrative and the Champagne region was publicly recognized as a dynamic part of the national territory.

It was probably no coincidence that fraud became a serious issue in the region with the harvest of 1900. This is the date generally recognized as the beginning of the mévente. Marked by a combination of surplus produce and low prices, the mévente was widespread, touching nearly all

French wine producers. "The succession of three years of abundance and quality in 1898, 1899, and 1900, the augmentation of the area planted in vines, [and] the use of prolific vine varietals," the Syndicat du commerce explained, "have effectively led to white wine production much higher than the needs of champagne manufacturers: it is to this principal cause that we must attribute diminishing prices and the mévente."[17] In the commune of Vertus, where vignerons had confronted authorities over the anti-phylloxera syndicate less than a decade earlier, it was estimated that production had jumped from a pre-phylloxera average of 28 hectoliters per hectare to over 38 hectoliters per hectare between 1898 and 1900.[18] The large supply of grapes meant "a quantity of wine sufficient for provisioning the entire world," the marquis de Polignac of the Veuve Clicquot firm joyously proclaimed. For the unfortunate vignerons, recovering from the economic slump of the 1880s and the phylloxera crisis of the 1890s, however, it meant surplus produce and falling prices.[19]

From the meager harvest of the phylloxera years, the situation had changed drastically and suddenly. Chamber of Commerce records indicate that there were enormous stocks of unsold grapes throughout the vineyards by 1901.[20] Few within the industry would have denied that there was a serious *crise viticole* developing in the vineyards of Champagne. In the newspapers and lecture halls of Paris, economists such as Charles Gide and Paul Leroy-Beaulieu fiercely debated the causes of this mévente.[21] But for the intransigent growers of common wine and the equally intransigent growers of grapes for champagne, the source of their misery needed no debate: it was a sea of surplus production made deeper by the rising tide of fraudulent wine—wine from vineyards outside the region—entering the market.

"In examining the question," wrote one vigneron from the Marne, "we see that this mévente has several causes: the principal cause is fraud of champagne wines."[22] Petitions from vigneron villages identified fraud as the chief worry of the rural population. The commune of Verzy, far from the center of the vigneron militancy of the 1890s, for example, sent a petition to the departmental administration asking that the "public powers intervene in order to curb the growing fraud of using the label 'vin de champagne' on wines that do not originate from the crus of Champagne." From the commune of Hautvillers, the home of the legendary Dom Pérignon, vignerons demanded that action be taken immediately against the "abusive employment of the name *champagne* made by the négociants and the fraud that is practiced to the great detriment of champenois interests."[23] Believing that they needed to protect "the reputation of our excellent wines," vignerons demanded that the "public powers repress with all their vigor the fraud that discourages vignerons already sorely tried by so many diverse calamities."[24]

Problems mounted in the fall of 1902 and spring of 1903. "The year 1902," recorded the Chamber of Commerce, "will count among the worst that Champagne has experienced." Having stocked their cellars during the abundant harvests of 1899 through 1900, négociants had little need for the equally abundant grape harvests of the following years.[25] "I do not need to describe the state of our vineyards," declared one observer; "everywhere, it is dire financial straits while awaiting complete ruin."[26] There was a startling contrast between the "dire financial difficulty" of the vignerons and the growing wealth of the négociants.[27] By July 1903, négociants reported a jump in sales of nearly 3 million bottles from 1902, reaching record high sales of over 31 million bottles of champagne worldwide.[28] Sales were booming, particularly in the domestic market, where consumers demanded millions of bottles of champagne each year.[29] With an increase in sales by almost 50 percent between 1900 and 1910, images of economic hardship and claims that there was a *crise viticole* seemed incompatible with the prosperity of the champagne firms.

Looking back nostalgically at the period before the mévente, Alphonse Perrin, a vigneron leader, reminisced: "Twenty years ago everyone wanted to be a [Champagne] vigneron, he had everything . . . he was always sure of his affairs, [grape] prices were always higher than he imagined, and the vigneron, happy and content, saw no clouds on the horizon." Examining "the causes that have brought such a deplorable change in the existence of the vignerons," Perrin concluded that phylloxera and the mévente were responsible for the economic woes of the community. But, he continued, hidden behind the immediate difficulties of the mévente were the real sources of economic dislocation: fraud and the Syndicat du commerce. If only the vignerons had listened to Lamarre and rallied together to fight fraud, Perrin lamented, they could have avoided their slow "strangulation." Perrin embraced Lamarre's idea of a "delimited Champagne." Fraud could only be stopped when vignerons were in possession of a law that delimited the boundaries of the area of grape production that could legitimately be used in champagne production by regional firms.[30] In short, Perrin wanted to see terroir made into a legally defined area.

Lamarre's biggest miscalculation, according to Perrin, was his belief that vignerons could solve their long-term economic difficulties through formation of a cooperative. "I want something more practical; in a word, I want something reasonable," he stated. With the appearance of the Syndicat du commerce, everything had changed for the vignerons, much like the introduction of "an electrical current." Suddenly, grape prices had dropped, and negotiations appeared to cease between négociants and vignerons. If the Syndicat du commerce was not responsible for fixing prices, then why had the situation changed, he demanded. "But after all of this, I do not reproach messieurs les négociants for joining together," Perrin continued. "It is their right." Vignerons, however, had an equal "right" to unite to negotiate grape prices with the négociants. The solution for vignerons when faced with two powerful adversaries, fraud and the Syndicat du commerce, was a "force for opposing another force." They needed the collective action of a "syndicat des vignerons," Perrin concluded.[31]

Although the impassioned rhetoric of the 1890s was gone, the lessons of the phylloxera struggle were not. The experience of the 1890s had proven the effectiveness of vigneron solidarity and collective action when backed by the power of the state. Unlike the previous decade, there were already local vigneron-based organizations in place. Most communes had one or more organizations that grouped together the majority of growers.[32] The hundreds of small anti-phylloxera syndicats united in the Association viticole champenoise provided a convincing example of the potential of collective efforts, while the networks of solidarity created during the struggle against phylloxera provided important links among vignerons across the Marne. While the AVC was a successful organization, it was structured to obtain government subsidies and coordinate phylloxera treatment and was thus unlikely to become a central body for rallying vignerons to a dual campaign to negotiate grape prices and fight against fraud. What was needed, mused Perrin, was a new, more powerful organization built on the networks formed by the AVC. What was needed was action to establish the boundaries of a vigneron-defined terroir.

Action came with the founding of the Fédération des syndicats viticoles de la Champagne (hereafter the Fédération), officially registered with the government in August 1904. Based on a single syndicat in the town of Venteuil, however, the Fédération was active as early as 1901, working to "protect for our wine the proud name of champagne."[33] By 1904, its membership included 102 local syndicats, nearly the same number of local syndicats belonging to the AVC.[34] Although there are no figures for the total number of participants, we do know that, organized much like the AVC, the Fédération was open to both vignerons and vigneronnes.³⁵ But, unlike the AVC, the Fédération also included small shopkeepers and rural artisans who were linked to vignerons through the village economy. The Chamber of Commerce noted that the local merchant "always suffers during bad harvests." With economic hardship among the vignerons, the small merchants and artisans of local villages found it increasingly difficult to survive. Moreover, these individuals were linked to the vineyards as Champenois, sharing the characteristics of a rural terroir that produced their collective esprit as well as the regional wines. The series of bad years meant that by 1903 "[village commercial] transactions have become more and more difficult as a result of the lack of money and the precarious situation of the vignerons."³⁶ Both the economy and the collective "soul" of the region cried out for change.

The importance of the solidarity between nonvignerons and vignerons in local villages was clear with the election of a master tailor from the village of Verzenay, Edmond Bin, to lead the Fédération in 1904. Although Bin himself was not a vigneron, he clearly identified himself and his fellow artisans and merchants as a principal part of the vigneron community. This community did not share the same interests as urban labor, regardless of links to champagne production. Critics blasted what they saw as an enormous contradiction in the discourse and practice of the syndicat. The role of nonvignerons in the organization was particularly incomprehensible. Mocking the lack of vigneron presence among so-called vigneron representatives, one critic remarked that "No vignerons, for the vignerons" should be the Fédération's motto.³⁷ Bin was criticized for leading a *syndicat mixte* that failed to include those intimately connected with the industry, including skilled wine cellar workers in Reims. Bin's response was clear: "What common interests, vignerons, could you have with the *cavistes* and *tonneliers*, who are not vignerons themselves?"³⁸

Only those outside of the Fédération found Bin's statement ironic. Vignerons and artisans in the Fédération, regardless of whether they owned vines or combined property ownership with wage labor, did not identify their interests with those of urban workers. Central to the identity of the Fédération and its membership was neither a link to the manufactured product nor a link of class interests, labor against capital. The link was terroir. In the attempt to transform "peasants into Frenchmen," so eloquently described by Eugen Weber, the term *terroir* reflected the changing relationship of the center and the periphery. By the end of the Belle Époque, while retaining its specific relationship to wine, *terroir* had come to designate a rural area considered as the primary basis of the collective character traits of the local inhabitants. Since 1870, however, in the interest of creating Frenchmen, the state had, according to Weber, attempted to eliminate these characteristics in the name of a larger, national unity.

The issue of fraud—and the crusade to protect the vin gaulois—opened the possibility of negotiating the meaning of Frenchness and the place of rural France within the nation. Terroir, for those in rural France, created the esprit gaulois; it was through its connection to the rural world that the esprit was transmitted. France could not be French without this link. For Bin and his followers, this link made perfect sense. For those intent on turning peasants into Frenchmen or defining rural France in terms of labor and capital, the discourse of Perrin, Bin, and the rural Fédération reflected the backwardness, the lack of collective, national consciousness of the rural world. Peasants appeared to conform to the classic model outlined by Karl Marx:

Each individual peasant family . . . acquires its means of life more through an exchange with nature than in intercourse with society. A small holding, the peasant and his family; beside it another small holding, another peasant and another family. A few score of these constitute a village, and a few score villages constitute a department. Thus the great mass of the French nation is formed by the simple addition of homonymous magnitudes, much as potatoes in a sack form a sack of potatoes.[39]

The vignerons, however, had a very different vision of how the "great mass of the French nation" was formed.

Far removed from the language of class conflict adopted by Lamarre and other militant vignerons in the 1890s, Bin and the Fédération adopted a tone of social cooperation, much like that of Léon Bourgeois, the tenacious Radical who was consistently reelected in the Marne between 1888 and 1923. Given the bloc of rural support for Bourgeois and the Radical party in the Marne, it should perhaps not be surprising that the rural membership of the Fédération echoed (and was echoed in) the attitudes and aims expressed by their political leadership. Bourgeois, the author of *Solidarité*, a work published in 1895, central to the doctrine of the Radical party, advocated a policy of social solidarity.[40] Solidarity was to be achieved by the state and voluntary organizations working to create institutions to protect the individual, "nominally set free, but all too often deprived of the means of effectively exercising this freedom." The Radical party, the pivot of political life of the Third Republic, provided the rallying point for those who, like Perrin and Bin, were calling for regulation of the industry and were "prepared to use the power of the state and encourage the activity of the trade unions and co-operatives . . . to transform a political democracy into a social democracy."[41]

Just as Bourgeois hoped to prevent the polarization of French politics into hostile camps, Bin appeared to want to avoid the division of Champagne into opposing groups. Just as Bourgeois and other Radicals turned away from a combative *revanchard* nationalism that continually looked toward the Vosges, the Fédération and the vignerons rejected the *revanchiste* language of the 1890s that linked négociants to German overlords. The Fédération stressed the strength of voluntary association, whether among vignerons or négociants, and the interdependence of both groups. In encounters with the Syndicat du commerce, Bin and the Fédération attempted to reach an entente, using polite, conciliatory language and stressing their shared interest in the prosperity of the wine industry.[42]

Bin approached the Syndicat du commerce with a request to negotiate grape prices first in 1901 and then again before the following three harvests. The Syndicat's response was clear: in no way would it "serve as an intermediary between the propriétaires-vignerons and the maisons of commerce" in what it considered purely private matters.[43] Moreover, the Syndicat repeatedly insisted, it had nothing to do with the prices established in the vineyards.[44] While Bin could not persuade the Syndicat to enter into price negotiations, the tone of its response was, nevertheless, cordial. In the first exchange, the president of the Syndicat cautiously agreed with Bin that both organizations could benefit by working together.[45] A series of increasingly friendly letters, expressing a shared interest in protecting the appellation of champagne, were exchanged between Bin and members of the Syndicat between 1901 and 1903.[46] Jointly, the vignerons and négociants negotiated a reduction of rail rates for workers coming for the harvest in 1902 and 1903.[47] These congenial but cautious first encounters between the two organizations opened the way for earnest discussions to protect the regional appellation, the common denominator that linked the vignerons of the Fédération and the négociants of the Syndicat.

While members of the Syndicat du commerce and the president of the Fédération were exchanging friendly letters, there were no collective efforts for recognition of appellations. The Syndicat had all but abandoned hope that there would be an international agreement protecting regional appellations from "false indications of origin." Passage of the Convention of Madrid in 1892 did little to stem the use of "champagne" as a generic label.[48] Only France, Great Britain, Spain, Switzerland, Portugal, Brazil, and Guatemala signed the final agreement, leaving some of the largest markets for regional sparkling wine unprotected. Altering its strategy, the Syndicat rallied other members of the fine wine industry to lobby the French government for comprehensive legislation recognizing and restricting wines using an appellation d'origine. A domestic consumer protection law to safeguard the appellation, it was reasoned, would both reassure consumers abroad of the "purity" of French wines and serve as the basis for arguing for a new trade agreement.[49] With little political clout in Paris and with attention focused on the growing political crisis over the Dreyfus affair, the Syndicat and fine wine interests found their two proposals for regulating appellations introduced in the French legislature but quickly abandoned.[50]

By the spring of 1903, the responsibility for studying the matter was left in the hands of the department. When discussion of the use of "grapes or wines foreign to Champagne for the creation of sparkling wines" began in the Conseil général of the Marne, the impetus came not from the négociants but from the rural population. Inasmuch as the vignerons had supported Bourgeois and other candidates of the left-wing Bloc des gauches coalition in the elections of 1902, their grievances attracted the attention of elected officials. Mounting stacks of petitions from vignerons that collected in the office of the prefect of the Marne made it clear that counterfeit champagne was a potentially volatile issue among rural voters.[51] The Fédération's president had exchanged letters with the president of the Syndicat du commerce in 1902 about laws that would delimit the boundaries of the department of the Marne and regulate the movement of outside wines within it. Syndicat members rejected these proposals, claiming that the question was better left to the courts, using existing laws, and self-regulation by the négociants.[52]

The creation in 1903 by the Conseil général of a commission for studying fraud forced the debate between the private forces of the Fédération and the Syndicat du commerce into the public sphere. During the sessions, however, it was not Bin, as president of the Fédération, who debated Paul Krug, president of the Syndicat. Rather, a local doctor, A. Péchadre, a member of the Conseil général with political affiliations to the Radicals, took up the position of the vignerons and the Fédération.

Evidence suggests that although not formally a member of the Fédération, Péchadre was actively engaged before this meeting in discussions of fraud and the mévente with the Fédération's leadership.[53] Politically ambitious, Péchadre emerged as a passionate advocate of the vignerons' call for reform. Elected to the Chamber of Deputies in 1906, he declared: "I have always been and I remain with the undeservedly exploited vignerons; I have always been and I remain against the fraud that they encounter; I have always been and I remain a loyal collaborator with the Fédération des syndicats viticoles de Champagne."[54]

In the opening session of the commission in 1903, Péchadre stated, "*Eh, bien* . . . in spite of the enormous costs that weigh on him, the vigneron champenois must again defend himself; he has never had a more dreadful enemy." It was less than a decade since phylloxera had made its way to the Marne, and fractious debates brought armed confrontations, so the specter raised by Péchadre of facing another "more dreadful enemy" in the midst of the mévente was ominous. "I mean," explained Péchadre, "the competition waged against vignerons by the introduction of outside grapes . . . in Champagne, substituting them for the grapes of the region in part of our sparkling wines." Echoing the language of the vignerons' petitions, Péchadre was unequivocal: this "dreaded enemy" was inside the region and had to be rooted out. If outside wines or grapes were used because of a poor harvest, there might be some justification, he exclaimed. But that was not the case in Champagne: "It is even in the years of abundance that this fraud is practiced."[55]

Péchadre proposed a system for legally delimiting the boundaries of Champagne and regulating the movement of outside wines within the Marne that was strikingly similar to the one originally proposed by Bin in his letters to the Syndicat du commerce. At the center of Péchadre's proposal was independent monitoring of champagne production within the Marne and the issuing of "stamps of authenticity" that would certify that the final product was "pure."[56] Experiments with a similar system of monitoring in the Charente for cognac production appeared to be working smoothly. Yet Péchadre anticipated objections. The négociants would complain, he argued, because they feared that "a light veil of suspicion might hang over the authenticity of their products." The interests of the champagne houses and the vignerons, however, were the same: protection of the reputation of champagne. United, he concluded, the champagne community "would be a good example of the solidarity between capital and labor."[57]

Péchadre's proposed delimitation of Champagne and controls on wine production went much further than previous legislation supported by the Syndicat du commerce. Objections came from the Syndicat at a meeting of the Conseil général in August 1903. Syndicat president Paul Krug agreed that there was a need to protect the regional appellation, but at the same time, he expressed Syndicat objections to any specific legislation that would bind "the honest" regional merchant-manufacturers. Krug argued that the Syndicat had always considered the boundaries of Champagne *viticole* to include simply the arrondissements of Châlons-sur-Marne, Épernay, and Reims, but the courts had ruled that the boundaries of Champagne included all of the departments of the Marne, Aube, Haute-Marne, and Ardennes. While the Syndicat preferred the smaller definition of "Champagne," he continued, it was not the place of the Syndicat or local authorities to obstruct "commercial liberty" by fixing geographic limits for purchase of raw materials (grapes) or to attempt to regulate the movement of wines within the confines of the Marne.[58]

In a reversal of previous syndicate statements, Krug noted that there might be some regional négociants, particularly those making less expensive wines, who purchased grapes and wines from outside of Champagne. This was unfortunate, he added, since Champagne was one region in France where the interests of the "négociant are intimately tied to those of the proprietor." But for the sake of both the négociants and vignerons, the Syndicat "considers the best guarantee of authenticity for the public to be the venerable reputation of their houses and the interests that they [the négociants] have in maintaining the quality of their wine and the prestige of their brands." With consumer preference for the authentic brands, he concluded, the law of supply and demand would eliminate these less exacting producers and further the "struggle against the competition of wines from outside Champagne." There was no reason to intrude on the "liberty of commerce," the right of honest producers to protect their trade secrets, when the market had the means to correct itself.[59]

Péchadre rebuffed the Syndicat du commerce because it appeared that it wanted the name "champagne" protected for the limited use of merchants within the strict confines of the Marne but was unwilling formally to link regional sparkling wines or bind merchants to purchase grapes from that same delimited area. If the interests of vignerons and négociants were "intimately tied," Péchadre asked of Krug before the Conseil général, then why did he approach the question of regulating wine production solely from the point of view of the négociants? "What preoccupies him [Krug] is the merchant," an exasperated Péchadre exclaimed, "and as for the unfortunate vignerons, there is no remedy. It is necessary, he says, to wait. But wait for what? That the law of supply and demand will come to reestablish the equilibrium and make them [vignerons] feel its benevolent influence in a time more or less distant?" All of the talk of commercial liberty simply made possible the "free exercise of fraud." It was urgent, he implored his colleagues, "to undertake a genuine national effort to stop the disappearance of the most French of all the wines of France."[60] As during the phylloxera debate, protecting champagne was equated with protecting the nation.

Péchadre's passionate denunciation of the Syndicat du commerce and dramatic call to fight against the "dreaded enemy" in the name of the French nation resulted in a unanimous vote by the Conseil général to send his proposal to the minister of commerce. It was hoped that the government would attach an amendment for protection of *appellations régionales de provenance des produits viticoles* to a law on fraud in the sale of wine that was currently under consideration in the Assembly. Assuming that delimiting the boundaries of Champagne would be essential to the application of any appellation legislation, a special committee, which included Péchadre and his colleagues, was created by the prefect of the Marne to craft further recommendations for delimiting the area within which legitimate champagne makers could buy their grapes.

Asked by the prefect to participate in the commission, Krug fired off a letter of protest. Avoiding direct reference to arguments raised during the Conseil général meeting, Krug explained that the Syndicat du commerce had been examining the question of fraud for years and realized that the issue was complex. Any open or public discussion by a special committee, he asserted, would be against the collective interests of the region. Krug warned that outsiders would use the committee reports "as propaganda for outside wines that usurp *our denomination of vin de champagne*" (emphasis in the original). Reasserting négociants' claims to be the protectors of the economic legacy of Champagne, Krug attempted to put a halt to the proceedings. Commercial liberty and professional secrets were at stake.[61]

The prefect could not ignore the groundswell of rural support for Péchadre and the special commission. By October 1903, he had received letters from over 163 communes supporting the commission's efforts.[62] Faced with the polarization of relations between vignerons and négociants, the Fédération used the pages of its newspaper, *La Champagne viti-*

cole, to reassert its desire for amiable relations with the Syndicat du commerce. In one issue, the editors applauded the possibility of cooperation between négociants and vignerons through the nomination of two members from both the Fédération and the Syndicat to the proposed commission.[63] Despite their differences, both the Fédération and the Syndicat had a stake in successful delimitation.

The article in *La Champagne viticole* indicates that the Fédération was willing to compromise with the Syndicat. But members of the Conseil général appeared much less willing to cooperate with négociants. Marcel Pochet, a vigneron member of the Conseil, declared that he had received petitions from vignerons in 110 different communes supporting efforts to stop "négociants buying wines from outside of Champagne." Vignerons were willing to support the protection of the appellation "champagne" for the use of local merchant-manufacturers, but only if these same négociants were legally bound to purchase grapes from the terroir controlled by local grape producers. With vignerons demanding regulation of the négociants, Péchadre declared, the committee had to move forward to assure that wine marketed as champagne was made exclusively from grapes grown by Champenois in a delimited area.[64] Protecting the place of manufacture was not enough. Péchadre hammered out an agreement with both the Fédération and the Syndicat to set the boundaries of Champagne *viticole* as the three arrondissements of Épernay, Reims, and Châlons-sur-Marne.[65] The ancien régime's province of Champagne, centered on Troyes, would henceforth be only a memory. A new Champagne was born. Members of the special commission brought the proposals to protect the freshly delimited Champagne to meetings with Bourgeois and other regional representatives in the French legislature.

Vignerons in the Fédération, négociants in the Syndicat du commerce, and elected officials from the Marne appeared, for once, united in their commitment to set the limits of the "true" Champagne. This was to be the end of counterfeit use of the name. The debate over the highly anticipated Law of 1905, however, showed how hard it was to define a product appellation. Describing the ingredients of a product, even of complex blended wines, should have been fairly straightforward. In the tradition of Pasteur, the disciplined science of winemaking established in France in the 1880s focused on discovering the constituent elements of wine. The grape varietals that went into champagne were well known. But a deeper understanding of the relationship between grape and wine remained more art than science. The science of enology was still in its infancy. Precise knowledge of grape maturity and how it determined the character of the resulting wine, not to mention vine genetics, were almost unknown until after World War II. The transformation of grape juice into wine required a few fundamental steps, but there were enormous variations in techniques, even among the makers of fine wine within the same region. Despite the fact that there were many trained wine bureaucrats in the Ministry of Agriculture, science offered few precise rules for establishing appellations.

Appellations required more than simply designation of grape varietals. Because much of French thinking about wines until this period focused on the relationship of soil and grapes to the final finished wine, the move toward appellations required matching the characteristics of wines with, at minimum, a territory or, at best, a terroir. As Raphaël Bonnedame had indicated when he attempted to define "champagne" in 1899, this was not an unambiguous matter. Not surprisingly, perhaps, it was far from clear how to proceed. Would it be best to create appellation laws and then, as had been the case in the past, allow the courts to decide the limits of their application? Would it be best to define the appellation by first defining or delimiting the region? But, in a modern France that appeared to be effacing regions and regional identities, how would a region be defined? Was this a matter of science? Economics? History? Or was wine subject to some other, transcendent "truth," as the Bordeaux classification system of the 1850s suggested?

These were questions that were not isolated within the discourse of rural France. By the turn of the century, the new science of geography, so closely linked with the ideas of Paul Vidal de la Blache and, later, the historical project of Ernest Lavisse, focused attention on the relationship of France's regions to the country as a whole. Through the works of these great thinkers, French intellectuals and elites were introduced to the "harmonious ensemble" of France that was created by balancing the diversity of its regions.[66] The school of French geography centered around Vidal and his faithful students at the École normale supérieure and the Sorbonne urged the observation of France's natural regions, not to retrieve forgotten traditions but to discover the harmony of the nation's existence. Henri Bergson, Léon Blum, Édouard Herriot, Jean Jaurès, Charles Peguy, and Romain Rolland were among the thousands who attended Vidal's legendary lectures.[67] "I discovered a passion for geography as it was revealed to me during the ... lectures of Vidal de la Blache," Rolland would write. "One saw *la Terre* as a large animal, a living organism ... it was *la Terre* that lived, that thought, that reacted, in us and for us."[68]

Central to Vidal de la Blache's classic works, most notably his *Tableau de la géographie de France* (1903), is the view that environment, shaped by internal and external factors, determines the way of life (*genre de vie*) of a locality and its people. According to this logic, a country can only be understood through its environment as determined by its geomorphology. France's geography, according to Vidal, situated it at the crossroads of Europe, the crossroads of "the civilized peoples."[69] France was more than the historical synthesis of all these civilizations, it was the culmination of civilization. Vidal's formulation of France, one historian has noted, "clearly owed something to nongeographical considerations."[70] Breaking away from popular nineteenth-century ideas that linked climate to a theory of natural borders, Vidal argued that France was a unity because of its "personality." This French personality, he argued, was a result of its diversity. Within the small area that was France, he noted, geomorphological evolution had created the diverse conditions that interacted to create France. It was man, however, who ultimately established the link between the disparate features produced by nature. The diversity of people within France, reflected in regional identities, gave the nation a unique ability to assimilate and "transform what it received."[71] France was an organic entity with a diversity of related parts, all of which supposedly gave the nation its unique personality.

While other nineteenth-century writers like Jules Michelet would argue that France's personality emanated from Paris, Vidal and his followers emphasized the importance of rural France, the France of *pays*. Throughout his body of work, Vidal gave a great deal of attention to detailing local life. He emphasized local names for regions and types of landscape, commenting on the intimate connection between the land and people. Often he contrasted these regional appellations with the sterility of generic labels created by politicians or scholars. Vidal looked to the soil and its inhabitants in search of France's personality, the roots of France's genius. Throughout his works, there are almost lyrical passages describing the rich relations between place and population. Indeed, the real life of France was the world of the people of rural France, who "embodied the *genius loci* that laid the groundwork for our national existence."[72]

Neither Vidal nor his followers attempted to find any scientific explanation or "necessary principle" explicit in nature for the existence of France. For the founder of French geography, there was no doubt that France was a gift, a land of treasures. "It is the abundance of 'goods of the earth,' as old folks say, that for them is identical with the name of France," he wrote at the beginning of the *Tableau*. "For the German, Germany is above all an ethnic idea. What the Frenchman sees in France, as his homesickness shows when he is away, is the bounty of the earth and the pleasure of living on it."[73] This indissoluble relation between the French and their natural world was eternal for Vidal.

The work of French geographers like Vidal de la Blache was adopted by the republican government of the Third Republic as part of its standard pedagogy for generations of French schoolchildren. While the geography texts of Vidal and his followers did not explicitly link agricultural products with the French personality, the elevation of the rural world to the center of a transcendent French civilization created a new prestige for regional products emanating from the soil. Wine, in particular, with its importance as the national *boisson hygiénique* was a material manifestation of the holistic relationship between the people and the natural environment. Indeed, statements found in popular scientific works show a belief in an animating spirit shared by people and wine.

These same lessons and this same passion for the "magic of the landscape" of France were passed on to thousands of French schoolchildren through the *Tableau géographique* and other Vidal de la Blache geography texts that were adopted in the schools under the Third Republic. Vidal taught that there was a natural symbiotic relationship in France between the people and the land. This was most clearly to be seen in the provinces, where diverse regional personalities emanating from the natural environment were still to be found. The visible natural diversity of the provinces was the immediate objective manifestation of the collective French soul. France, for Vidal, was nothing without its regions. And wine was one of the "principle agricultural assets" that linked France's soil and soul.

If the science of geography demonstrated the harmony of France's regions, it also revealed the natural, fundamental divisions within France. In his pathbreaking essay "Des divisions fondamentales du sol français," published in 1888, Vidal had forcefully argued for a new administrative and territorial division of France that had "more of a relation with contemporary realities."[74] He believed that the centralization of modern France was excessive and had to be reversed by creating new boundaries more closely related to "new economic needs" and social realities.[75] This theme, repeated in his lectures, reproduced in different forms in his later works, and taken up by many of his followers, found its way into the debate over appellations and delimitations. Indeed, Vidal's geography text *La France*, in its fifteenth edition by the end of World War I, specifically addressed what he called, "the true champagne" and the value of French wines, saved from the phylloxera by science and "the patience, the sacrifices of the viticulteurs."[76] Although Vidal was not directly involved in the appellations and delimitations discussions in 1905, his ideas about the value of France's regional personalities and natural divisions, such as that of the "true champagne," were widely consumed by the political elite and the general public. It was the scientific nature of these arguments that gave them their power.

By 1900, there appeared to be a popular climate that was more or less open to a reconsideration of the value of the region. A discussion of the place of the region in the nation had imperceptibly entered into the popular imagination. One example of this comes from the immensely popular *Le Tour de la France par deux enfants*, a classic children's story by Augustine Fouillée (published under the pseudonym "G. Bruno") about two "grown-up" boys who venture alone across France.[77] Historians have studied the travels of the two main characters of *Le Tour de la France*, André and Julien, to understand the formation of collective identity at the turn of the century. Intended as a reader for French schoolchildren, the story of the excursion of André and Julien sought to demonstrate that regional differences in France were "gifts to the nation," which came together to create a unique "French temperament."[78] By the time the appellations debate had become national, several generations of French men and women had mastered this primer in school. Readers were familiar with the notion that all the regions of France were linked, trading products as well as their intrinsic qualities.

There were constant reminders of the regions' values. "I must protest against the new trend of selling, under the name of French wines, products that are not from our terroir," the world-renowned chef Auguste Escoffier wrote in 1906. "It is truly derogatory, for the excellent reputation of our crus, that such an inaccuracy should be tolerated."[79] Having made a career of promoting French cuisine and building menus around French regional specialties, Escoffier, like many of his counterparts, was a staunch supporter of the need to protect the wines of France.

A glance at the menus created for clients such as Edward VII, Kaiser Wilhelm, and Sarah Bernhardt at the Petit Moulin in Paris or the Ritz of London demonstrates the chefs' appreciation of French wines and, particularly, champagne. An elite clientele came to expect not only chilled bottles of Veuve Clicquot and Moët et Chandon, however, but also carefully chosen regional specialties such as "peaches of Montreuil," "butter d'Isigny," "asparagus of Argenteuil," and "veal from Pauillac."[80] For Escoffier, like many French taste professionals, regional terroirs created unique tastes, which were featured on menus and worthy of international recognition and protection. As Rebecca Spang has pointed out in her work on the invention of the restaurant, *la carte* in French means not only a menu but also a map.[81] The practice of listing regional specialties on *la carte*—whether in a restaurant or in the textbooks of French schoolchildren—reinforced the links between the science and the art of cooking and the nation.[82]

But the problems at the heart of the appellations and delimitations discussion were not as simple as those presented in children's books or on restaurant menus. France risked internal divisions—protection of some regions and regional producers would come at the expense of others. Given the complexity of the issues and the enormous pressure from rural France to "fix" the market, the final law created in 1905 was, from the beginning, completely inadequate. The legislation itself was designed as a consumer protection law. Legal action could only be undertaken when "fraud" relative to the origin of the product was intended to deceive the consumer. Consumers, not the grape growers who supplied the raw materials, would be the injured party and have to police the industry through the courts.[83] The idea was, essentially, that informed consumers would make rational choices.

During parliamentary debates concerning the Law of 1905, the French sparkling wine producers of Champagne led the charge, but they were not alone. Other elite producers joined in the campaign for delimitations. They were also joined, however, by other sparkling wine producers, particularly those of the Loire, who also lobbied hard for inclusion in the appellation laws.[84] The result was an impasse. Arguing that appellations and delimitations were the only way to end the *crise viticole*, representatives of the Marne introduced an amendment that would grant appellations protection but leave the question of delimiting territory to the minister of agriculture.[85] While clearing the way for the passage of the 1905 law, the amendment simply shifted responsibility for sorting out conflicting claims and mediating the divisive debate without actually solving any of the concerns of angry vignerons. A consumer protection law with no mechanism for regulation, making it difficult to prove or pursue fraud, had not been the goal. But, in the eyes of vignerons of champagne, the most serious failing of the legislation was that the actual delimitation of wine regions was left to the discretion of the minister of agriculture. Protecting wine appellations was opening larger questions about what, at base, was a region in the new modern, capitalist France. These were questions, however, that went well beyond the wine industry itself.

The day following the promulgation of the Law of 1905, the minister of agriculture called on the Conseil d'État for help in determining the origins of wines, designating appellations, and coordinating these with delimited territory. The Conseil d'État refused, arguing that the Law of 1905 remained ambiguous on the question of delimitations and, thus, they had no reason to proceed. Once again, delimitations brought the nation to an impasse. In the Marne, the Conseil général and Fédération put extensive pressure on the local administration, the minister of agriculture, and Marne representatives for "complementary measures" to specify areas to be delimited as the source of specified wines. Many outside the industry regarded the idea of delimiting Champagne and regulating the industry as senseless. In a biting article in *Le Temps*, the Fédération was attacked for using the issue of fraud and delimitations to reignite old tensions within the region. There was no *crise viticole* in Champagne, the author exclaimed, and, in comparison to the Midi, the vignerons of the region were thriving. The public, the article concluded, was tired of the complaints of those who made a fortune from sparkling wine.[86] Economists, too, scathingly criticized the protectionist nature of potential delimitations and the interruption of free markets within the nation.

There were those within the Marne who were having second thoughts about the need for state-enforced delimitations. With the unrelenting push of the Fédération and members of the Conseil général for regulation, the Syndicat du commerce pulled back. In an extraordinary session of the organization in 1906, the négociants in the Syndicat agreed to institute a system of self-regulation, adopting several suggestions made by Péchadre. The organization would institute a system of colored tickets that would accompany all wine shipments within the Marne. Only wines made from tickets verifying that the wine was from the region would be allowed to use the regional appellation. All other wines, it was decided, would simply have to be labeled *vins mousseux*. This system of self-regulation was consistent with earlier Syndicat regulations and the general attitude of the organization throughout its history.

In a letter that accompanied the proposals to the minister of agriculture, the Syndicat stated that the boundaries of "Champagne" had to remain flexible "to permit all houses loyally manufacturing *vins à bon marché*" to supply themselves with grapes. Contradicting their previous agreement with the Fédération and members of the Conseil général, the Syndicat proposed enlarging the boundaries of Champagne.[87] Despite this clear reversal, the Fédération, under the leadership of Bin, continued its moderate stance vis-à-vis the négociants, attempting to compromise with the Syndicat in order to maintain their fragile solidarity in the face of vocal opposition. The Fédération circulated petitions to protest the proposed enlarging of the boundaries of Champagne in April 1906.[88] A month later, however, the leadership reversed itself.[89]

At the general meeting of the Fédération that was held in March 1906, Bin, sitting next to Raoul Chandon, a representative of the Syndicat du commerce, announced the decision to compromise with it.[90] The decision to modify the Fédération's position, ignoring the signatures of vignerons collected on petitions only one month earlier, appears to have brought dissension within the vigneron community. At the meeting on March 27, 1906, where Bin made his announcement, Émile Moreau, a Fédération member from Aÿ, presented the "grievances of the small proprietors, always sacrificed for the large, and who are never paid for their labor." The small proprietors would never find a voice within the Fédération, Moreau believed. Admonishing them to create a separate organization, he argued that "it is only a union of all the small proprietors that can resist capitalist power."[91]

In many ways, Moreau's response demonstrated the extent to which the position of the vignerons was changing rapidly in these early years of the twentieth century. With the price of grapes plummeting and costs rising, these small proprietors were increasingly reliant on wage labor. Those forced to sell their vines had wage labor as their only recourse. Thousands moved into the cellars as *cavistes* in Épernay, Reims, and other smaller towns; thousands of others found wage labor on large estates. Yet the rationalization of production and increase in labor supply had the adverse affect of reducing wages.[92] Moreau, along with his brother Jules, attempted to organize both a small producers' cooperative called the Syndicat commercial viticole champenois for these small proprietors and an association of vine workers and small proprietors called the Syndicat des ouvriers champenois.[93] A poster produced by the cooperative is reproduced in figure 17. The Syndicat des ouvriers champenois had sixty members in 1906, mainly in Lamarre's home base of Damery and around Épernay.[94] Similar organizations were founded in neighboring communes although they often did not include *ouvriers vignerons* as part of the organization's name until after 1911.[95] Although these organizations did not present a serious alternative to the Fédération until after the riots of 1911, they challenged the position of the Fédération's leadership throughout the intervening years and highlighted the rapidly shifting economic position of the vine-growing community.

Without access to Fédération archives, it is difficult to discern the extent of dissension within the ranks of the organization in 1906. Although there was no apparent shift in the Fédération's relations with the négociants, there were a number of dissenting voices among its leaders. Veiled exchanges in the pages of *La Champagne viticole* throughout 1906 and 1907 indicate there was an opposition faction within the leadership, organized mainly around Alphonse Perrin, Victor Philbert, and Gaston Poittevine. There does not seem to have been any suggestion like Moreau's, however, that the Fédération should split into separate organizations. In an indirect attack on the leadership of the organization, Perrin, in a small pamphlet on the progress of the anti-fraud syndicats, argued that a central organization was needed to take a more aggressive stance with the Syndicat du commerce.[96] The position of these three men must have had some support within the organization, because they emerged as its leaders after Bin's resignation in 1909.

Increased purchases of the grape harvests in Champagne during 1904, 1905, and 1906 by the négociants ameliorated the situation in local villages, perhaps temporarily easing the indignation of vignerons faced with continued reports of counterfeiting in the Marne.[97] But the position of the vignerons was still precarious. "The vignerons are waiting anxiously for a law that will delimit the actual borders of Champagne *viticole*," a Chamber of Commerce report noted in 1906; "they hope that this measure will facilitate the sale of their products and will relieve the downward spiral of prices.... If that law does not produce the desired effect, disillusion will be total and the end result disastrous."[98] With desperate southern growers demonstrating in the streets of Narbonne, and the Midi on the point of exploding, there was an increased sense of anxiety in Champagne vineyards.

In some communes, growers tried to regain control over the situation. In the prized grape-producing area of Aÿ, for example, growers created an association to press their grapes communally and certify the authenticity of the wines produced.[99] Other angry vignerons from Champagne bombarded the minister of agriculture with over 100 petitions demanding a public hearing on fraud in Champagne.[100] Finally, in June 1907, the Chamber of Deputies authorized meetings to proceed with delimiting wine regions, based on *les usages constants*—whatever these were determined to be. Under pressure from Léon Bourgeois and Péchadre, recently elected to the Chamber by rural voters from the Marne, the minister of agriculture agreed to public hearings on fraud and the possibility of delimiting Champagne based on "customs, which, because of their age, their universality, and their morality, have the authority of a law."[101] Only the regions, as Vidal de la Blache had taught, could decide these types of issues.

While some, like Vidal, had great faith in the regions, others saw the issue of delimitations as little more than a destruction of the nation, a potential civil war. In one mocking article in the pages of *Le Figaro*, the delimitations discussion was titled "La Guerre des Deux Haricots," or "The War of the Two Beans," and depicted as a national calamity. The satire began with a description of hostilities in the provinces of France, dubbed the Kingdom of Little Peas. Two Beans, each originating from a different region, confronted each other at the market. One Bean argued that he was the superior vegetable, representative of the refined riches of the kingdom, endowed with "unique qualities" and heir to a rich historical legacy. His opposing legume, in the outlandish dialogue that followed, attacked these assertions by laying claim to very similar "unique qualities."

Mounting anger turned the Beans an unnatural shade of green. On the verge of violence, the seething Beans called out to the Peas, who legislated for the kingdom from their seats in the National Pod. Each Bean demanded special status within the kingdom derived from his intrinsic superiority. The Peas, for their part, found themselves sharply divided over the claims of the recalcitrant Beans. Incensed by the other's rival claims and frustrated by the inaction of the Peas, the two Beans savagely attacked each other.

Fighting quickly spread. The noisy battle between the two Beans attracted the attention of other members of the Kingdom of the Little Peas. A truffle from Périgueux, a stick of butter from Isigny, an asparagus stalk from Argenteuil, and a wedge of cheese from Brie vaulted into the raging conflict. Amid the cacophony of competing claims, other vegetables, cheeses, and wines joined in demanding special legislation to acknowledge their distinctive qualities. War escalated. As they oozed and bruised, splattered and spewed, each belligerent clamored louder for state intervention on its behalf. The Peas, paralyzed in their Pod, watched with horror as the "The War of the Two Beans" tragically engulfed the kingdom.[102]

"The War of the Two Beans" would have made a wonderful folktale explaining the origins of a particularly rich and delectable French dish. But the editors of *Le Figaro* had less light-hearted goals behind publishing this fable. The grotesque "war story" was a cutting critique of the mounting demands emanating from the provinces of France for legal protection of agricultural products at the beginning of the twentieth century. Debates over the parameters of the legislation and its potential international implications were closely followed throughout the wine-drinking world. The final outcome of the "War of the Two Beans" had both widespread and long-term implications.

Producers of champagne, ridiculed as the beans who had launched the war, argued that their reputations and their national heritage were being usurped by those who treated local product appellations as generic labels. Their ruralist and protectionist discourse elevated the land and its products to a "central part of every Frenchman's legacy."[103] Indeed, Pierre Nora notes in the introduction to a monumental study of French memory, that "the 'land' [in France] would not exist as such without the ruralist and protectionist movement" of this period.[104] Throughout the Belle Époque, contention over controlling the land and its legacy through geographically based names seemed poised to degenerate into a "War of the Two Beans." The satirical food fight presented in *Le Figaro* seemed at least as much prophecy as parody.

For those who produced France's sparkling wines, their claims to protection were not as petty as the satire in *Le Figaro* would have had readers believe. Boundaries were critical for wine. It may seem self-evident that wines from Algeria were not from the Bordeaux region of France. Labeling Algerian wine as "bordeaux" would, thus, mislead the consumer. But what about the vineyards outside of the key villages or core production centers? The town of Blaye, just opposite of Margaux, is one example. Historically, it had produced white wines for distillation into cognac, but it also produced robust red wines, which were periodically marketed as bordeaux. While the classifications from 1855 did not include these growths, they did not exclude them from benefiting from the more general appellation of bordeaux. In this case, fixing the Gironde River as a boundary would cause considerable social disruption for the wine communities outside of the boundary of demarcation.

The problem was more acute for complex blended wines such as champagne. Champagne was made from wines from several years and many vineyards. What was to be the final boundaries of the vineyards when the product bore only a general place-name? And what about wines made outside of the wine region? Could wine produced in Germany with grapes purchased exclusively in the primary vineyards near Reims, for example, legally be labeled "champagne"? There were, indeed, a number of German firms that regularly bought French grapes to make sparkling wine. And there were even a number of French firms that had production facilities in Germany in order to avoid customs duties and potential losses of bottled wine in transport.[105] The more expensive and exclusive the wine, the higher the stakes were in delimiting boundaries.

It was the effort at delimitation that brought the relationship between the French people and their soil into the center of the discussion. The concept of *appellations d'origine* was based on the principle that finite geographical boundaries existed beyond which wines and those who produced them were no longer entitled to use a regional appellation. Geographic delimitations were used to legitimate and limit access to the national patrimony. Different groups within Champagne sought to promote their positions, their concerns, as those of the nation, demanding a national response. State efforts to appease the various "protectors" of the patrimony by delimiting the Champagne region launched the "War of the Two Beans."

Readers of *Le Figaro* had little difficulty identifying the two recalcitrant beans in the satirical war. The wine-producing community of the Marne and the less numerous but equally vocal vine growers of the Aube were the two beans poised for battle. After receiving hundreds of petitions from angry peasant vine cultivators in both departments, the minister of agriculture was charged with the establishment of a commission to examine delimiting the vineyards. A public hearing on the issue was held in May 1908. Hearings were to determine and demarcate the boundaries of Champagne based on local use of the name, historic production techniques, and natural factors such as soil and climate. The actual meeting, however, turned into a battle over the construction of the French past.[106]

The public gathering was as tumultuous—although not as messy—as the *Figaro* satire suggests. Official representatives of the peasant vine cultivators and merchant-manufacturers of the Marne outnumbered representatives from the Aube, the Haute Marne, and the Aisne. Marne supporters packed the hearing room despite protests from the delegates from the other three departments.[107] Marne representatives arrived with clear objectives: they wanted legally recognized boundaries that corresponded to a limited area of grape cultivation, all of which fell within the department of the Marne. Other areas of vine cultivation—and the peasant cultivators of those areas—were to be excluded. Grapes grown outside a carefully defined area would not be Champagne grapes and thus could not be used for the production of sparkling wines bearing the regional appellation. Aube representatives, in particular, seethed with anger over the exclusion of their grapes.[108]

The transcripts from the public hearing reveal how each of the "Beans" attempted to demonstrate their superiority by claiming to be the heir to a rich historical legacy. Speaking for Aube wine producers, Deputy Thierry Delanoue opened the hearings by challenging the historic construction of "champagne." The city of Troyes, he began, was the historic center of Champagne. Holding the meeting in Châlons-sur-Marne, he argued, was a move to promote the interest of a few producers who wanted to disfigure the limits of Champagne by creating a "Champagne without a capital, completely deformed, where it is necessary to abolish all the old memories that hold together the territory." Perhaps more damning, he argued, the new Champagne, in a delimited territory with a commodity-based boundary, would be detached from the Champagne of the people, the Champenois. To support this, he produced a series of petitions and protests signed by Champenois from his department.[109]

Delanoue went further, questioning the "paternity" of champagne by proclaiming Clairvaux, a town within the confines of Aube, to be its "birthplace." Dom Pérignon was not the legitimate "father" of champagne. Monks from a monastery in the Aube, he claimed, had been the first to cultivate the grape varietal known as "d'Arbanne." Wines from these grapes were later used at the abbey of Dom Pérignon at Hautvillers in the Marne to produce sparkling wine. Champagne had, indeed, been invented by a monk, but it had had many fathers. The Aube could rightfully claim status as the birthplace of champagne and the legitimate inheritor of the fathers' legacy. No descendent of this heritage could claim, he argued, to be more champenois than anyone else.[110]

Questioning their legitimacy as heirs to Dom Pérignon was almost too much for the citizens of the Marne. Hearing transcripts note the tense tone of the Marne representatives in the aftermath of Delanoue's remarks. Marne growers turned to their representatives to launch a counterattack. The facts regarding the origins of sparkling wine were well established and well known, Marne representatives argued in quick succession. Dom Pierre Pérignon, a saintly blind monk, was said to have used his highly developed senses of smell and taste to develop sparkling wines at the Abbey of Hautvillers between 1668 and 1715. He was thus considered the "father" of champagne, the inventor of sparkling wine, and the founder of the champagne industry. "We [growers of the Marne] remain the inheritors of not only the name 'champagne,' but of the vines that still produce the same celebrated wines," the Marne vignerons' representative Michel-Lacacheur argued in a rhetorical flourish.[111] They had never strayed from their ancestral legacy.[112]

Not only did the Marne representatives claim to be the true heirs to the ancien régime foundations of the industry, they also claimed in the public hearings to be the valiant protectors (along with the republican leaders of the Third Republic) of the legacy of the French Revolution. Calling for small vineyard boundaries, representatives of the Marne argued that the Revolution had justly dismantled the ancient province of the counts of Champagne. The feudal boundaries of the Champagne region, which had once bound the peoples of the Aube and Marne, had been abolished in 1790. In place of the oppressive imposed boundaries of the ancien régime, a new, organic Champagne *viticole* had been created based on winegrowing traditions and a productive relationship. This was the area that was, by right, the Champagne region. Enlarged boundaries advocated by vine growers in the Aube, they argued, were little more than a thinly veiled attempt to reestablish feudal boundaries, recreating the feudal privilege of the nobility. The Aube, they concluded, was attempting to reverse the progress of the Revolution.[113]

The French past, and the place of the regions within that past, was constructed very differently by representatives of the Aube. Countering those who advocated strict boundaries, Aube representatives argued that "champagne" was a national patrimony, a wine that linked them as a people. The proposed limited Champagne region was deformed, they argued, because it was not representative of a democratic France. A new, modern Champagne would be exclusively based on the economic relations of a small group within the department of the Marne. Champagne, they argued, belonged to all Frenchmen as part of their heritage. By legally demarcating the boundaries of the region within which it could be made, the representatives from the Aube concluded, the Third Republic would reestablish the worst of the ancient régime privileges. "The Revolution united France," avowed one representative, "abolishing the ancient provinces; we do not want to reestablish them with their privileges, their prerogatives, their barriers; we must remain faithful to our principles of Justice and to Liberty."[114]

What is interesting for the historian in this exchange is that both sides attempted to appropriate the French Revolution, but did so by revising the standard view of the impact of the French Revolution on the provinces. Alexis de Tocqueville, writing in 1856, noted that there were those who viewed the ancien régime as the golden age of provincial liberties, which had been crudely violated by the Revolution.[115] Those who believed in a pre-1789 provincial "golden age" were often discredited as separatists or monarchists by republicans who championed a vision of France as a centralized, unitary republic. Representatives of both the Marne and the Aube, using the rhetoric of the Radical republicans who dominated Third Republic France, argued that the Revolution had, in fact, released the provinces from oppression. But while the Aube representatives argued that no new regional boundaries should be established, the Marne representatives presented a vision of a postrevolutionary, republican France with new, organic regions. The "father" of the industry might be connected to the ancien régime, but the regions themselves were at the core of a republican France.

Tempers flared as these competing claims were set forth. Physical confrontation between the "two beans" was averted by the abrupt decision by leaders of the Aube delegation to abandon the hearings altogether. Overwhelmed by the sheer number of representatives from the Marne and contesting the legitimacy of the proceedings, members of the Aube delegation declared that there was little point in voting on the commission's final recommendations and walked out as an act of protest. Less than an hour later, the commission passed a resolution recommending to the minister of agriculture that the boundaries of Champagne be determined to consist of viticultural areas found exclusively in the Marne department, with the addition of a number of communes in the department of the Aisne (thirteen in the canton of Château-Thierry and twenty-one in the canton of Condé-en-Brie).[116]

🍇

The exchange was noisy but hardly a war. Tensions mounted following the hearing, but the Syndicat du commerce and the Fédération continued to present a united front during the wait to receive approval of the proposed delimitations from the Conseil d'État.[117] On December 17, 1908, the cabinet accepted the decision of the commission, granting Champagne the first legally recognized regional delimitation. Other fine wine growers demanded similar proceedings, and delimitations were quickly granted to Cognac, Armagnac, Banyuls, and Bordeaux.

On the eve of the harvest of 1908, government opposition to delimitations of Champagne would have undoubtedly brought armed confrontation in the Marne. The parallels between the economic plight of growers in the Midi and those in Champagne were evident in headlines that read, "After the Midi, Misery Visits Champagne."[118] It was clear in July of that year that the harvest would be a disaster for the vignerons: mildew and other vine diseases had nearly wiped out the entire grape harvest.[119] Local officials called on the government for emergency aid.[120] Vignerons' impatience for a system to regulate fraud, according to one reporter, had turned into exasperation. "At this moment 32 casks of outside wines, destined to replace the destroyed harvest, are expected . . . the arrival of these convoys could bring an explosion."[121]

Posters calling vignerons to a meeting in Aÿ were found plastered on village walls throughout the Marne in September 1908. The anonymous author entreated vignerons to present their grievances directly to the Fédération's president. "Now more than ever we must unite to defend our interests BY ALL MEANS NECESSARY. . . . The moment to act has arrived," the poster implored in bold letters.[122] The meeting, in the courtyard of a school in Aÿ on September 13, 1908, drew a crowd of nearly 2,000 of the region's 15,000 vignerons. Angry vignerons complained that négociants, anticipating the passage of the delimitation decree in December, were doubling the movement of outside wines into the region.[123] Jules Moreau demanded that the vignerons begin an immediate tax strike to protest delays in passing delimitations. Tempers flared in the crowd; voices were raised to demand a special agent to immediately investigate fraud. Others advocated violence, yelling: "It is too late. Outside wines are entering our department! We must empty these barrels!" With the escalating rhetoric and some minor scuffles in the crowd, Péchadre attempted to soothe the vignerons. While offering a long apology for governmental delays, he implored the vignerons to support the Fédération and avoid violence.[124]

The meeting in Aÿ demonstrated the level of vignerons' frustration by the end of 1908. Anger was directed as much against merchants bringing in outside wines as against the Fédération under the moderate leadership of Bin. Annoyed at Bin's conspicuous absence at their meeting, vignerons demanded explanations from the vice president about the continued collaboration between négociants and the Fédération. Of particular concern was the continued accord between the Fédération and the Syndicat du commerce, despite years of refusal on the part of négociants to negotiate grape prices with the vignerons.[125] Michel-Lechacheur, one of the directors of the Fédération, said that if the vignerons did not agree with the decisions of the leadership, they would willingly resign. Vignerons greeted this announcement with wild applause.[126]

Vigneron leaders such as Perrin who had previously veiled their criticism of Fédération policy now openly accused the Syndicat du commerce of price fixing and the négociants of fraudulent production.[127] Just as quickly as Bin could come up with conciliatory letters to the Syndicat assuring it that the vignerons of the Fédération desired continued cooperation,[128] angry vignerons made new accusations.[129] Refusal by the Syndicat to meet with the Fédération to discuss grape pricing engendered bitterness. "In the vineyards," wrote one observer, "the intransigence of the négociants provokes a great deal of discontent."[130] Breaking the accord with the Fédération, vignerons reported to *Le Matin* that the only solution to the economic crisis of Champagne was to delimit the region, stop the importation of wine from outside, and fight the Syndicat du commerce with a strong, vigneron-based organization.[131]

In a damning series of articles in *Réveil de la Marne*, a négociant admitted buying wines from outside the region and discussing grape prices with other négociants before the harvest.[132] Publication of these statements prompted a flurry of responses by négociants. The Syndicat du commerce briefly rebutted that prices were fixed by the simple law of supply and demand.[133] But the strongest response came from E. Morinerie, a négociant outside the organization and president of the Reims Chamber of Commerce. Under Morinerie's leadership, a pamphlet was issued protesting the accusations made against the négociants and the Syndicat. Referring to the "abominable press campaign," he wrote that all the accusations were purely political. Although he did not specify the objectives or the source of these political attacks, he made it clear that the Syndicat and nonmember négociants were united in their objections. "The Syndicat does not consist of ALL the houses," he asserted, making it impossible for négociants to fix prices.[134]

Referring specifically to the articles in *Le Matin*, Morinerie accused the vignerons of "ingratitude" after the efforts made by the négociants of the Syndicat du commerce on their behalf during the phylloxera crisis and the funds of négociants placed at their disposal through the AVC. The négociants had attempted to aid the vignerons, but still the latter blamed them for the fall in grape prices. "In an earlier period," he continued, "many vignerons, particularly in the *crus secondaires*, made red wines that they sold as *vins de table*, and found in this commerce a satisfactory remuneration for their efforts." Seduced by the high prices paid by the champagne houses, he argued, vignerons had foolishly abandoned this market for red wines, and, in their greed, selling to négociants "soon became their only objective." Echoing the words of the Syndicat, he argued that the law of supply and demand dictated the market. If the vignerons abandoned alternate markets, then they must accept responsibility for their growing economic woes.[135]

While négociants appeared united in the face of accusations of fraud and price fixing, the growing dissent among the Fédération membership threatened to split the vigneron community. Prompted from within the leadership, there was a shift in control of the Fédération. Vice President Victor Philbert, who had led the veiled criticism of Bin, now replaced him at the helm of the organization. Gaston Poittevin assumed control of the organization's newspaper, and Perrin became its secretary. With increasing tensions in the vineyards, the leadership faced the dual battle of preserving unity among the vignerons and obtaining some measure to enforce the delimitations granted by the minister of agriculture in 1908. Seeking the "strength" that Perin had counseled in 1907, the Fédération announced that it would affiliate with a new organization, the Confédération générale agricole pour la défense des produits purs, which was led by an individual identified simply as Bolo-Pacha.[136] Little is known about this obscure businessman from Marseille, who allegedly was made a pasha by the khedive of Egypt. Bolo entered the wine industry through his marriage to a young widow who owned a successful wine house in Bordeaux. During World War I, he was arrested for espionage, reportedly having used his government connections to spy on France for the Germans, and was shot for treason at Vincennes on April 17, 1918.

Much like Marcelin Albert in the Midi—a little-known café owner who emerged as the "redeemer" of the desperate vignerons of southern France during the mass protests of 1906–7— Bolo-Pacha was dubbed the "savior" of the Champagne vignerons. His program for saving Champagne was never clear. In many ways, his absence from the region and vague calls for "pure" products made him more of a symbol of the sense of grievance among vignerons than a leader of a regional vigneron movement. Vignerons writing in *La Champagne viticole* implored Bolo to join them in the vineyards during the despair that accompanied the meager harvests of 1909 and 1910.[137] What the Confédération générale agricole pour la défense des produits purs seemed to offer in 1909 was an organization for countering lobbying efforts directed against the concept of a Champagne *délimitée* and a leader to act as a symbol for vignerons' grievances.

By May 1909, the Fédération indeed seemed to have acquired a new "force." President Philbert obtained permission from the prefect of the Marne and the Conseil général for members of the organization to act as agents to monitor fraud.[138] The Fédération had a tool for pursuing fraud, but the Syndicat du commerce refused to agree to outside monitoring until "the day the denomination of 'champagne' is protected in some manner that is truly effective."[139] A bitter dispute began between the Syndicat and the Fédération over vignerons' competence in matters of fraud.[140] This came amid the continued spread of mildew and a subsequent drop in grapes supplies in 1908, 1909, and 1910.

Undaunted, the Fédération's agents went forward, monitoring wine in Champagne *délimitée* and pursuing legal proceedings against several small négociants and their firms in the courts of Reims and Épernay.[141] Representatives from the Marne, fearing unrest like that experienced in the Midi, attached a law concerning regulation—the long-awaited complementary measures—to the budget bill of 1910. This action appeared to appease the rural population until, under pressure from the liquor retailers (Syndicat de débitants de boissons), the Chamber removed the articles, stating that they would be taken up at a later date.[142]

Less than five months later, the situation was declared critical in Champagne: terrible storms washed away soil and vines in the hills around Épernay; mildew and insects were attacking vines with a vengeance. "Many vignerons are abandoning their vines," the Chamber of Commerce stated, "refusing new expenditures, which appear useless in the presence of a seeming disaster."[143] "For months," a writer in *La Champagne viticole* declared in August 1910, "we had confidence in the promises that were made to us. The few impediments to the *fraudeurs*' flourishing business do not appear sufficient; from every direction, [someone] announces a huge shipment of outside wines into Champagne to replace our destroyed harvest. Yes or no, will the laws be applied?"[144] Legislation had to be passed quickly to stem fraud, Poittevin warned: "Vignerons [are] in such a state of exasperation that a simple word would suffice to make them act outside the law."[145]

Tensions escalated with the disastrous harvest of 1910. Denouncing the Fédération for its inertia, Jules Moreau and 550 angry vignerons met in Damery on October 9, 1910, to organize a tax strike. Taxes, levied on the land instead of the harvest, were a repeated complaint in the Marne, where land values were high and incomes from the harvest erratic. Within months, vignerons throughout the valley of the Marne joined Moreau. The Fédération quickly announced a general meeting of vignerons in the *Réveil de la Marne*.[146] Between 10,000 and 15,000 vignerons marched through the streets of Épernay to join the Fédération on October 16, 1910, to cries of "Down with fraud!"[147] Fraud was the order of the day, and a *cahier de doléances* was created by the angry vignerons, denouncing the ineffectiveness of the government of Paris, the incomplete laws created to deal with fraud, and the refusal of négociants to negotiate grape prices.[148] Given the level of frustration in the vineyards, the meeting was remarkably calm. The leaders of the Fédération appeared to be in control of the growing protest movement.[149]

Members of the Fédération brought their *cahiers de doléances* to Paris on November 15, 1910. Accompanied by a delegation that included the "savior" Bolo-Pacha, Senators Vallée and Montfeuillard, Deputy Péchadre (Deputy Lannes de Montebello of Reims, a member of the Montebello family of wine manufacturers, was absent), and the prefect of the Marne (Chapon), the delegation met with the president of the Conseil d'État to request that immediate action be taken on the long-awaited "complementary measures." Several weeks later, the text of the proposed law sent to the Chamber of Deputies was published: finally, specific regulations for monitoring fraud and use of the name "champagne" were clearly spelled out. Discussion of the measures, however, was not scheduled until the following month.[150]

The situation was becoming increasingly tense in the vineyards. Throughout the months of November and December 1910, the "agents" of the Fédération stepped up efforts to seek out and prosecute fraudulent producers in the "war against fraud." The press reported that angry vignerons were packing the courtrooms to assure "justice" in cases against fraudulent merchants.[151] One potentially volatile case of fraud, known as "L'Affaire Coste-Folcher," implicated numerous négociants in a web of deceit. Authorities feared further inflaming tempers and pushed back hearings on this case until January of the following year.[152] On November 8, the subprefect of Épernay was forced to intervene when nearly 400 vignerons from Damery intercepted a shipment of outside wines at the train station in Épernay. Under pressure from the crowd, the owners of the shipment were forced to send the wines back to the Midi. Violence was averted in the train yard, but that evening, windows were smashed and barrels of wines that remained at the station were emptied, "while the employees of the railroad were occupied elsewhere."[153] Poittevin, a leader of the Fédération, was able to calm protestors in Mesnil-sur-Oger a week later, but his intervention failed to stem the growing tax strike and a second protest in Damery on November 17.[154] By December, groups of twenty to thirty vignerons had been formed in Hautvillers and Dizy to act as self-appointed agents to scrutinize the movement of wines into the cellars of négociants in Épernay.

Paris felt the tension emanating from the vineyards of France. Economists weighed in overwhelmingly against delimitations because of their protectionist nature and their adverse effects on free trade.[155] Pundits, like the one who penned "The War of the Two Beans," were critical. There were, however, prominent intellectuals who supported the delimitations efforts. Important support was provided by the geographer Vidal de la Blache, who argued in the *Revue de Paris* in December 1910 that the departments that currently existed in France were artificial divisions resulting from compromises made during the heated days of the Revolution. The modern economic reality of France, changed by transportation and communication networks, required a new organization. Hinting at the debates dividing the nation, Vidal explained that it was the "obligation of the legislature to take into account all [the facts] and to create the best solution for the good of the collectivity." To help in this task, Vidal reminded the legislature that regions that produced coveted specialties tended to have manufacturing centers that served as focal points for economic activity. These centers were often the heart of natural regions.[156] In Paris, these "scientific" arguments of economics and geography became a central part of the debate.

Similar procedures for delimiting Cognac, Armagnac, Die, and Bordeaux now began. Tensions over who would be included and who would be excluded were acute. Passions were high, but there was no violence in these regions. Champagne became the exception. There, the vineyards seemed on the brink of conflict. Vignerons no longer appeared willing to wait for action. Delimitations and "complementary measures" were no longer simply a way of stopping the "dreaded enemy," as Péchadre had declared seven years earlier. Delimitations were seen by the desperate vignerons as a source of power, a force to be used to counter the social and economic power of négociants. While the Syndicat du commerce continued to oppose vigneron-initiated regulations, fraudulent wine appeared to be pouring into the department of the Marne. The debate over proposed geographic boundaries for Champagne had linked négociants and vignerons in a united community to protest the "pretensions of outsiders."[157] Their united front during these highly publicized meetings gave the impression of a harmonious community, united in a joint effort to protect their economic interests.

Indeed, in most narratives of the history of champagne, this period between 1900 and 1910 is invariably described as a period of harmony based on a common front against fraud, a mere interlude between the phylloxera "revolt" and "the revolution in Champagne" in 1911. Central to this interpretation is an assumption that the formative impulse for all demands for appellations and delimitations in the champagne industry came from the négociants. This was, not surprisingly, the view of the négociants themselves. "Our repeated complaints against the usurping of the name Champagne," Paul Krug stated in the annual report of the Syndicat in 1906, "has finally found an echo among the viticulteurs champenois."[158] Following this logic, the vignerons of Champagne were the "echo," the supporters of the Syndicat du commerce between 1900 and 1910. Attempting to explain the events that led up to the riots in Champagne in 1911, one historian has stated that the idea for delimitation had been raised by the négociants, but that "it was in the vineyards that application of [this idea was] achieved through incessant and energetic action."[159] While the vignerons were not passive, according to this view, they were, nonetheless, the docile supporters of their staunch allies the négociants.

At first glance, there appears to be little reason to question this interpretation. Statements made by négociants like Krug, combined with the amicable relations that developed between the Syndicat du commerce and the Fédération, point to exactly the kinds of conclusions reached in these standard accounts. The seemingly endless memos, technical reports, and legal studies generated between 1900 and 1910 give no hint of contention between groups within the industry. Moreover, this vision of a harmonious industry led by négociants is embraced and reproduced by those currently promoting the champagne industry.[160] Sifting through private correspondence and minutes from public debates, however, gives a very different picture, easily lost sight of amid the complicated legal and technical issues. What emerges is a picture not of harmony and consent but of conflict and compromise; a glimpse of a dynamic rural society struggling to define the boundaries of its community and the position of that community within the French nation.

Hidden was the friction of a community in transition. The fragile entente created by Bin and Krug a few years earlier could not withstand the resurfacing of the deeply rooted differences between négociants and vignerons, part of a long and recurring battle to define roles within the regional economy and the community. The contention and compromise that marked the period from 1900 and 1914 came from the inevitable adjustments to the changed social and economic position of members of the community in the face of new market conditions. To varying degrees, both groups called on the state to protect the economic legacy of champagne. Groups within the community, however, had different views of what form this protection should take based on their differing relationship to champagne—the wine, the land, and the business. Although the Fédération and the Syndicat du commerce had managed to mediate their differences, by the end of 1910, it was no longer clear that vignerons and négociants could find a compromise when faced with their versions of the "true" Champagne. Moreover, it was no longer clear that the nation itself knew what the place of regions, such as the "true" Champagne, was to be in modern, capitalist France.

CHAPTER

REVOLUTION AND
STALEMATE

The Revolt of 1911

"It is the final struggle in Champagne / Against fraudulent wines! / Vignerons of the Marne / Defend our wealth!" the vignerons of Damery sang to the tune of *L'Internationale* as they marched toward Cumières on the afternoon of January 17, 1911.[1] News had reached the vineyards that a shipment of 3,000 bottles of Midi wine was about to be delivered to the Perrier cellars. With drums beating, bugles blaring,[2] and blasts from *fusées paragrêles*, or hail rockets, word of the shipment quickly spread to the neighboring vineyards of Venteuil, Fleury, Binson-Orquigny, and Boursault. A red flag was raised in the center of Damery, and nearly 3,000 vignerons and vigneronnes moved purposefully toward Cumières.[3]

As the angry crowd approached the cellars, the frightened driver of the truck transporting the wine shipment backed into the interior courtyard of the Perrier maison. The outside gate was quickly secured. Within moments, the angry crowd had thrown open the gate, descending on the truck and its contents. A small group of protestors violently rocked the truck. The wheels began to roll. Together they pushed the truck nearly 200 meters to the edge of the Marne River. With one violent, collective thrust, the protestors rolled it over, dumping the entire contents into the water. Meanwhile, another group forced their way into the Perrier cellars. The owners, securely hidden away in the upper reaches of the building, would report hearing the unmistakable sound of thousands of bottles breaking. Three hours later, when the subprefect of Épernay and public prosecutor arrived to investigate, the shattered glass from 15,000 bottles of champagne, mixed with debris from barrels that once held 2,000 liters of wine, was all that was left in the cellar. Wine flowed across the grounds. The crowd had dispersed, leaving machines for bottling and materials for production strewn throughout the courtyard, most of it beyond repair. Remarkably, no one was injured.[4]

Finding the individuals responsible was difficult if not impossible. Reporting on the incident to the minister of interior, the prefect of the Marne noted that all inquiries were met with silence by the vignerons. "This was not an incident where protestors grouped in one locality or in one faubourg of a village," he explained, "but one in which individuals, scattered throughout the countryside, came in groups of three or four to a designated place, where they decided to take action."[5] Despite these obstacles, the following day, the public prosecutor amassed enough evidence to summon two vignerons from Venteuil for questioning. During the course of the long interrogation, a rock was thrown through the window, and, according to officials, they were surprised to discover that a "crowd, hostile and large," was gathered in front of the mayor's office where the vignerons were being held.[6] Tensions escalated as the crowd milled about awaiting news of the fate of the two vignerons. Concerned that an arrest would be provocative, the prefect ordered the release of the detainees. His intuition was correct. The ecstatic crowd embraced their comrades and peacefully dispersed with another round of exploding *fusées paragrêles*.[7]

With the release of the vignerons, the vineyards gave the appearance of calm. Police reports indicate that the tensions remained high. Sometime that night, the doors of the cellars of Berthet of Hautvillers were forced open and fifty-four half-barrels of wine were emptied. In the hours before the incident, rumors circulated that this wine had been made from grapes from outside the region and was destined for champagne production. Word spread throughout the vineyards that the owner planned on moving the wine to an unspecified location within the next two days. It was allegedly to be turned into champagne under the protection of troops from Épernay.[8] Many villagers heard the rumors; but no one witnessed (or admitted to witnessing) the destruction of the cellars.[9] The police suspected several groups of vignerons, but no one was arrested. According to police informants, members of this group were overheard saying, "Prison does not matter much to us; blood must flow. We shall go that far if necessary."[10]

These moments of violence in January 1911 were brief, but over the next six months, the conflict escalated into what many would later call the "revolution in Champagne." By April 1911, vignerons and soldiers were clashing in the streets and vineyards of Champagne. Property damage from burning, looting, and pillaging was extensive, and the costs were estimated in millions of francs.[11] The violence was, in retrospect, not surprising. Vignerons declared themselves "ready to kill or be killed for the cause."[12] During the six months of conflict and nine months of military occupation of the Marne, they amply demonstrated how far they would go. Pictures of the tense faces of the vignerons against the backdrop of smoking ruins of some of the nation's most prestigious champagne cellars seemed visually to capture the cause of the "revolution in Champagne"— class conflict.[13] Indeed, Jean Jaurès, the undisputed leader among French socialists, wrote that events in Champagne indicated the state of class relations in France. For him and many who followed him, the revolt proved that peasants were ready to join the struggle against capital alongside the proletariat.[14] The cause was class struggle; a struggle of labor against capital.

Historians who have studied the events of 1911 have confirmed the importance of class-based antagonisms in the conflict. Undoubtedly, notions of class did motivate some of those who participated. Symbols of the Left were clearly visible, and songs of class struggle were sung. But notions of a struggle of labor against capital alone are not sufficient to explain the actions of the rural population. Peasants—whether property owners or wage laborers or both—were united against négociants. But this unity was only within a clearly defined area of Champagne *viticole*; peasants outside this area who hoped to sell grapes for champagne were equally seen as the enemy, intent on fraudulent production. And not even all the négociants were enemies. Within the Marne, the pillaging and burning of négociant-owned cellars and vineyards was not indiscriminate or universal. Action seemed deliberate and selective. Some manufacturers were clearly and consciously exempted from the wrath of the roaming crowds. And throughout the tense months of conflict, the négociants themselves were split between those who supported "the cause" of the vignerons (although not the methods) and those who did not. The police undoubtedly played a role in the outcome of the conflict by blocking access to some of the key wine production areas and thus protecting some négociants. Nonetheless, there is clear evidence that vignerons selected their targets based on "the cause," which was not at base class struggle.

The cause that propelled them into the streets in 1911 was quite simple: to protect the connection between "champagne," the wine, and "Champagne," the land and its terroir. The government of the Third Republic appeared to agree that the link between the region and the wine needed to be preserved, passing a series of measures in 1905 and 1908 that would limit the use of the appellation "champagne." Public hearings in 1908 established the exclusive use of the appellation "champagne" for the vine growers and winemakers of the Marne, who successfully argued that the feudal boundaries of the province of Champagne had been abolished by the Revolution. Progress since the Revolution had created a new organic Champagne *viticole*, much like that described by Vidal de la Blache, based on winegrowing traditions and a productive market relationship. This modern region and its products were threatened. Legal measures to delimit the boundaries within which grapes used in wines labeled champagne could be grown were useless, the vignerons had discovered, without a system for monitoring or enforcing the legislation. Négociants continued to make "champagne" from grapes grown in neighboring departments and even the Midi, often to the detriment of Marne vignerons. Moreover, there appeared to be continued debate over the exclusion of grapes grown by peasants in the neighboring departments of the Aube and Aisne, included in the prerevolutionary province of Champagne. Legislative debate ultimately recognized the unique place of regional wines in the nation when Champagne was delimited in 1908, and again in February 1911, when the Bordeaux region was officially delimited. But the government of the Third Republic remained uncertain about the limits of champagne, both the wine and the region. For the rural population, the French nation existed to protect what was organic and authentic; the natural regions popularized by the French educational system and geographers such as Vidal de la Blache were the backbone of all that was essentially French, and the state was not living up to its commitment to them.

By 1911, vignerons were losing faith in the Radical-led government. Gaston Poittevin, the leader of the main vigneron organization, the Fédération, warned that the vignerons were "in such a state of exasperation that a simple word would suffice to make them act outside of the law."[15] Ironically, vignerons chose to "act outside of the law" by attempting to enforce the existing appellation laws. Taking the law into their own hands as we have seen, crowds of vignerons vandalized suspect wine shipments at the Perrier and Berthet cellars.

The cause for the vignerons was stopping fraud by forcing a definitive official recognition of the regions, particularly "Champagne," within the nation and providing the means for protecting the products of the regional terroir. Fraud could only be perpetrated, they believed, by dishonest négociants and grape growers who were "outsiders" with little connection to the terroir. Vignerons in the Marne no longer possessed the means to produce their own wines, having abandoned the costly production of sparkling wine to the regional négociants. They retained control of grape production, however, and an attachment to the soil, to the terroir, that was at the base of regional claims that the wines of Champagne deserved special protection through appellations laws. Lobbying efforts for protective legislation for wines of the Marne had presented champagne as the property of the French nation, carefully guarded by the community of producers in the Marne.[16] Champagne was singled out for special status among the wines of France, which, as Patricia Prestwich has noted, were already "encrusted with national myths about the glory and genius of the French race."[17] Both vignerons and négociants from the region argued that the appellation "champagne" represented not only a wine but also a community of producers that contributed to the glory of the French nation. The notion of fraudulent production, therefore, meant that there were those who threatened the community and, by extension, the nation. The cause in 1911 was thus a patriotic one.

When violence began in the Perrier and Berthot cellars in January 1911, there were premonitions that hostilities would escalate and become more widespread. Several négociants sent letters to the prefect of the Marne asking for troops to protect the transport of their wine. The prefect rejected these requests. Despite the vehement protests of négociants, the position of the prefect was understandable. An anonymous letter forwarded by the police to the departmental administration warned that "if anyone moves any wine, things could turn tragic."[18] Concerned about retaining the calm of the rural population while facilitating the free movement of goods, the administration finally responded with points of "special protection." Négociants concerned about their wines could bring the barrels or bottles to a designated area that would be protected by troops. With barely enough troops available to protect the Gare des marchandises de Reims, where several shipments of wine were awaiting transport, this was, négociants complained, a short-term, limited solution.[19] The vineyards were said to be in a state of siege.[20]

The government in Paris appeared well informed about tempers in the Marne, although it did nothing to stem the tide of protest. Besides the memos to the minister of the interior from the prefect, elected officials from the Marne were reportedly hearing from their constituencies.[21] Péchadre, a deputy from the Marne and self-proclaimed candidate of the vignerons, was allegedly informed about the protest movement by his uncle Marot, who was in contact with many of the vignerons' leaders.[22] Moreover, Lannes de Montebello, elected as the deputy for Reims with the support of its négociants, appeared to have contacts with both the Syndicat du commerce and the Montebello champagne firm.[23] Deputy Péchadre, echoed by Montebello, queried the government in Paris regarding complementary measures to enforce appellations. The issue of appellations and delimitations was once again open for reconsideration.[24] There was no indication, however, that the government would effectively deal with the issue to the satisfaction of the self-designated protectors of Champagne.

Patrols were stepped up in the vineyards as local villages prepared for the feast of Saint Vincent, which had been resurrected from a virtually forgotten religious festival to become a symbol of vigneron solidarity. At dances held the night of January 21, 1911, in Damery and Venteuil, soldiers sent to patrol the vineyards and vignerons fraternized until the early hours of the morning.[25] Similar fraternization had taken place in the Midi before the riots of 1907. "It is IMPOSSIBLE to foresee where we are headed," the subprefect of Reims noted ominously on the morning of January 22.[26] The Fédération had recently called for a tax strike and was now demanding a system of internal controls to assure that outside wines were not allowed into champagne blends. While Marne delegates worked in Paris to get a definitive law on delimitations, a series of meetings took place between departmental administrators, vignerons, and négociants in an attempt to diffuse the situation. In the town of Venteuil, the prefect of the Marne met at the Café Lemaire with 2,000 vigneron delegates from eighteen communes. In exchange for promises from vignerons that there would be no more sabotage of wine shipments, the prefect pledged that he would intervene to stop any transport of suspicious wines until the Assemblée nationale had voted on the complementary measures for delimitations. A similar meeting took place that same day between the subprefect of Reims and vigneron protestors and négociants in Aÿ and Reims. A written notice was posted in villages stating that an agreement had been reached between vignerons and négociants to cease wine shipments until complementary measures had been passed in Paris, in exchange for a cessation of violence and demonstrations.[27]

Over the next several days, Prefect Chapon met with delegates from among the négociants in Épernay and surrounding areas to explain the need to halt wine shipments.[28] Their agreement, however, seemed tentative at best: négociants throughout the Marne were impatient to resume wine shipments. Négociants only reluctantly agreed to await the passage of the additional measures scheduled for discussion in early February.[29] Within days of his meetings, Chapon was exasperated by the complaints and noncompliance of négociants. A delegation of "principal négociants" met with him on January 24, renouncing their promises to abstain from shipping wines until the vote in Paris. Over 700,000 liters of wine were stranded in the train station of Épernay alone, the delegates argued. When several threatened to send trucks to pick up their shipments, the prefect "informed the négociants that if . . . in spite of everything, they decide to effect such an operation, I shall decline all responsibility."[30] The following morning, representatives of the négociants and leaders of the Syndicat du commerce met again with the prefect, explaining that his refusal to give them protection to transport their wines was "causing great harm to all the merchants."[31]

Despite this apparent united front, however, the négociants were not all of the same mind. According to a police report, they fell into two groups: Reims négociants, who were willing to stick by their agreement with the prefect, and "the négociants of Épernay whose commerce includes wines from outside Champagne *délimitée*."[32] But the fissures among the négociants were not just a matter of geography; they were also divided by status. On the one hand, there were the firms outside the Syndicat du commerce, which were mostly small but included the former grande marque firm of Mercier. On the other, there were the grandes marques firms that belonged to the Syndicat, which tended to be grouped in the city of Reims.[33]

Another police report relating information about a February 3 meeting between factions indicated that négociants were divided into two well-defined contingents: "the first comprised the important négociants of the region, Polignac, Krug, Blondeau, Heidsieck, etc., and favorable to government projects [for delimitations]; the second comprised of the small tradesmen, Blum, Delouvin, Bourgeois, etc., and agitated by the same projects ... even tolerating the introduction of a certain quantity of wines from other vineyards of France in the manufacturing of champagne."[34] Whereas those in the first group listed in the police report were members of the Syndicat du commerce, those in the second were from among the leadership of a newly formed group called the Syndicat de défense des intérêts des négociants en vins de Champagne (hereafter Syndicat de défense).[35] The Syndicat de défense was led by Paul Fontaine, along with Eugène Bourgeois, "Père" Delouvin, and S. Blum, all of whom owned small champagne firms producing exclusively for the domestic market.[36] It claimed in its newspaper, *La Champagne commerciale*, published in Reims, to represent 180 small négociants who owned similar firms but were not included among the grandes marques of the Syndicat du commerce.[37] Many of these firms were in Épernay and the more peripheral areas of champagne manufacturing.

Efforts on behalf of the négociants of the Syndicat de défense were anonymous, often launched through the pages of its newspaper. According to articles in *La Champagne commerciale*, the real threat to the prosperity of Champagne was not fraud, but the *grandes maisons*. The delimitations, the newspaper proclaimed, would assure "a veritable TRUST or MONOPOLY OF OUR WINES to the profit of several grandes maisons and to the detriment of the majority of other houses."[38] The Syndicat du commerce, which represented only "34 houses of the 200 houses existing in Champagne," wanted complementary measures to assure this monopoly. Delimiting "champagne" would mean a jump in grape prices, it reasoned, the costs of which the small-to-medium-sized firm could not pass on to customers who preferred a low-priced wine.[39] Eventually, these smaller firms would be destroyed, allowing the Syndicat du commerce and the grandes maisons to control champagne production, grape prices, and the regional labor market as far as winemaking was concerned. With a perception that their economic survival was at stake, the négociants of the Syndicat de défense launched violent attacks against complementary measures, the entire system of delimiting Champagne, and the claims of the Syndicat du commerce to "speak" for the industry.[40]

The grandes maisons of the Syndicat du commerce were thus much closer to the position of the vignerons than to the stance adopted by their fellow négociants. Breaking ranks with the largest wine houses, the smaller firms of the Syndicat de défense demanded protection for the free movement of grapes and wine, a demand that was firmly rejected by the prefect.⁴¹ "Individual interests must be sacrificed for the general interest," he argued in a letter justifying his decision to the minister of the interior.⁴² The general interest in late January and early February of 1911 was preserving calm in the vineyards. Police reports on the state of the vineyards indicated that the majority of the rural population of the Marne supported the vignerons. Local railroad workers were "morally in agreement with the vignerons," one report noted. But that was not surprising, the investigator continued, given that "nine-tenths of the [local] population" supported the vignerons.⁴³ With the "moral support" of the railroad workers, vignerons were warned in advance about shipments of Midi wines destined for Épernay. Groups of vignerons, called together by the now-familiar blasts of *fusées paragrêles*, were able to intercept many of these shipments as they passed through the train station at Aÿ. Wines were either destroyed by the addition of gasoline to the barrels or the conductor was forced to return the shipment.⁴⁴ This sabotage appeared spontaneous; Fédération leaders were not implicated, although authorities were concerned about their leadership of the more generalized tax strike.⁴⁵

❦

The vignerons' struggle against fraud seemed less a class conflict than "an affair of the *monde viticole*" against those who threatened the "general interest" of the community.⁴⁶ The threat came not only from regional négociants who continued to use grapes and wines from outside of the region but also from the vignerons of these "foreign" vineyards who continued to sell their products to négociants in the Marne. Vignerons outside of the understood regional boundaries were equally the enemy. The intense debate in the Chambre des députés in February 1911 over complementary measures for delimitations that might include these other vignerons—particularly in the Aube and Haute-Marne—split deputies from various wine regions throughout France much the way it split the peasantry itself.⁴⁷ While no one in rural France contested the need to protect regional products, there remained enormous controversy over where the limits of those boundaries should be drawn.⁴⁸ The revisiting of the 1905 *zone d'appellation* law in 1908 and again in February 1911 indicates the continued uncertainly about the boundaries of modern regions within the nation. The final vote on February 10, 1911, for complementary measures reserved the appellation "champagne" exclusively for vignerons and négociants within the Marne. Eight days later, vignerons in Bordeaux—still working under the classification system of the Second Empire—were given official delimitations. Although greeted joyfully by those inside the delimited areas, these laws were immediately denounced in other areas of France.

Peasants in the Aube, the department neighboring the Marne, were particularly incensed at their exclusion. Since the phylloxera had appeared in the Aube in the 1890s, there had been a massive rural exodus and abandonment of vineyards, primarily those producing ordinary wines. Unlike in the Marne, there was no consolidation of large vineyards, and hence no expansion of wage labor. Those vignerons who remained struggled to hold on to their land by switching to quality grapes for sale directly to Marne champagne producers. By the early 1900s, this was remunerative, and grape growers relied on these annual sales. According to Chamber of Commerce documents, large quantities of Aube grapes were regularly purchased by champagne manufacturers, at least from 1888 to 1907. Their exclusion from Champagne *viticole* would mean economic ruin for their growers.

On February 26, 1911, the first acts of resistance began in the village of Fontaine in the Aube. By early March, a more generalized tax strike had begun throughout the grape-growing areas of the Aube. Red flags were hoisted as signs of defiance. In hopes of defusing the situation, a delegation of local officials met with the minister of agriculture, Jules Pams, in Paris, but it appeared that there was little desire to reopen the question. Between March 18 and March 24, 1911, a coordinating committee was created to organize collective efforts to overturn the delimitations, and local officials submitted their resignations in solidarity with the rural community. Police reports indicate that throughout the department, protestors showed their anger at the republic by painting over the motto "Liberty, Equality, Fraternity" on public buildings. Discontent appeared to be mounting.

Angry peasants of the Aube were joined by négociants of the Marne from the Syndicat de défense in opposing the boundaries set by the delimitation law. In *La Champagne commerciale*, a bitter attack was launched against the legislation, the Syndicat du commerce, and the representatives of the Marne in the Chambre des députés.[49] The principal argument against the delimitations became social rather than economic. With the reduction of the boundaries of Champagne, the small négociants would be forced out of business, the paper argued, creating a monopoly for big business. This, members of the Syndicat de défense contended, would cause widespread unemployment among cellar workers and other laborers employed by small champagne houses.[50] Although it did not call for unity between capital and labor, the rhetoric of the Syndicat de défense implied that the organization believed its solidarity was with peasants of the Aube and Aisne, workers, and small-scale merchants and manufacturers.

Believing that their future was seriously threatened by delimitations, peasants in the Aube and their representatives, along with the Syndicat de défense, sought to reverse the legislation. Angered by these attempts, the négociants of the Syndicat du commerce and vignerons from the Marne united. During the month of March, mass demonstrations and political action were organized throughout the northeast of France. On March 20, in the Aube, nearly 8,000 protestors took to the streets chanting "Injustice creates revolt!" and "La Champagne aux Champenois!"[51] The activities of the Syndicat de défense and attacks launched in *La Champagne commerciale* fueled peasant protest movements in the Aube. Although there is no direct evidence linking the Syndicat de défense to protestors in the Aube, Jean Nollevalle contends that "it is very possible that the négociants of Épernay financially contributed to the movements in the Aube."[52] An interview with Bertrand de Mun, president of the Syndicat du commerce, published in May 1911 and police reports that year seem to confirm this view, implicating a number of members of the Syndicat de défense in an exchange of funds with protest leaders in the Aube.[53]

Rural France was torn apart by the issue of delimitations. Class solidarities of the kind outlined by Jean Jaurès, which pitted labor against capital, had little relevance in the heated atmosphere. What was at stake was determining the boundaries of membership in Champagne. With membership came economic control over part of the national patrimony. Delimitations, however, meant an internal demarcation that would recognize some French citizens as protectors of the patrimony of champagne while excluding others. Peasants of the Aube, representatives from other wine regions in France,[54] and négociants of the Syndicat de défense recognized this and called for rescinding the law. For the vignerons of the Marne, however, defining them as Champenois, who alone could legally produce champagne, was a nonnegotiable duty of the Third Republic. The privileges of some regions had to be recognized. The "cause" was not strictly regionalist but one that demanded the recognition of regional identities and, if necessary, their protection. By March 1911, mass meetings were being held in the Marne to denounce the "inertia" of Paris in protecting the Marne peasants in the face of the "pretensions of the Aube."[55]

The government in Paris was deadlocked over the delimitations issue in March and April of 1911. The conflict went beyond the concerns of the wine industry. At the state level, the conflict revolved around defining the place of "regional personalities" within the modern nation-state. Representatives from the Marne, joined by those from other fine wine regions, argued that France's wine regions were part of the *patrie* deserving protection. Deputies and senators from other areas of France argued that "delimitations divided the country, created local animosities, and weakened national sentiment."[56] Much to the dismay of the representatives of the Third Republic, the periphery was not collapsing in the face of national integration. It was at once reinventing itself and demanding recognition of its place within the nation.

There seemed to be no easy solution to the impasse. Politicians were as divided as rural France on the question of delimitations. While there was a great deal of interest and activity by members of the Socialist party during the protests in the Aube, the events of early 1911 were far from "socialist in the manner of the great Jaurès."[57] With the dissension between regions and the threat of violence among the peasantry, "the socialists were left to follow the lead of the Radicals for solutions to the peasant question."[58] But the Radical party itself seemed divided on the question of delimitations and protection of agriculture, making it an unlikely leader in solving the dilemma fracturing its electoral base. A bill was proposed in the Chambre des députés to expand delimitations to include the Aube,[59] but members of the Senate proposed a separate bill to rescind all measures connected with appellations and delimitations.[60] Even the persuasive Marne representative Léon Bourgeois, who was elected president of the Chambre with Jaurès as vice president, could not rally Radicals and Socialists to consensus or compromise on the proposed bill. Debates in both chambers appeared as fractious as events in rural France.

As debate in Paris ground to a halt, peasants in the Aube and the Marne mobilized.[61] On April 8, nearly 350 vignerons took to the streets in Bar-sur-Aube, red flag flying, and were quickly joined by local factory workers. A socialist by the name of Checq was rushed to the front of the crowd as the "redeemer of Champagne." When the group reached the city of Troyes, police estimated that there were nearly 20,000 protestors proclaiming their right to the appellation "champagne." While peasants in the Aube were marching in Troyes, the angry leaders of the Fédération in the Marne turned the Café du Soleil in Épernay into a temporary headquarters.[62] In the Marne, posters with a skull and crossbones motif proclaiming "Death to Fraud!" and "Long live the true Champagne!" were plastered on walls. A group calling itself the Comité de défense des intérêts des vignerons ruinés de la Marne challenged the Fédération to take radical measures to assure the delimitations law. "'Be calm, be calm,' the Fédération tells us. But why?" one poster asked.[63] On Tuesday, April 11, the Fédération met with leaders from 102 of the organization's local syndicats, counseling patience while elected representatives continued their struggle in parliament.[64]

"Nous les vengerons! This is our battle cry!" an anonymous writer for the Comité de défense des intérets des vignerons ruinés declared in April 1911. Playing on the similarities between the words *vigneron*—grape grower—and *vengeron*—avenger—the writer called on vignerons to avenge the many victims of fraud within the region. Efforts would have to be systematic, the writer explained. To aid the vignerons in this task, the "enemies of Champagne" would be enumerated in a three-part hand-written leaflet sold by the Comité beginning on April 8, 1911.[65] While the archives leave little information on the composition of the Comité or the identity of the writer, police reports reveal that the leaflet was explosive.[66] Of particular interest to the vignerons (not to mention the police) was the "Livre noir des assassins de la Champagne" (Black Book of the Assassins of Champagne), a hand-written catalogue of the "falsifiers and the hereditary enemies of the poor vigneron."

According to the writer, much of the fraud was not openly perpetrated by the négociants, who had reputations to protect. Fraud was conducted with their knowledge, however, by their agents or representatives, who could work relatively undetected, or by other négociants, who sold their bottled wines to the larger houses for distribution. The first to experiment with fraudulent production, according to the pamphlet, was the wine house of Mercier, currently run by a German of the "Jewish race" who was not a naturalized French citizen. Mercier was joined in 1890 by another fraudulent producer, the writer continued, the firm of Moët et Chandon. Once these large producers actively engaged in fraud, "small and large *fraudeurs*," some likewise accused of being "foreigners," all jumped in "the same sack." The pamphlet continued by detailing the links between the various fraudulent producers, through agents or direct sales, and their precise methods of falsifying wines. Whether the charges were true or not is impossible to tell. Some of the charges—such as making wines from pears and apples—are highly suspect. The vehement hatred of Moët et Chandon and hints of a Jewish conspiracy undermine the argument even further. In the complex world of wine sales and the highly secretive tradition of blending, however, the detailed nature of the writer's charges and the ability to trace links between agents and wine houses remains impressive. The fourth section of the "Livre noir," to be issued after April 11, was to list, alphabetically, all regional fraudulent producers and their interactions.[67] Events in the Marne would explode, however, before the last installment was printed.

Circulation of the "Livre noir" was not the only concern of local officials. Apprehension was also growing over the spontaneous movement of support for the vignerons of Champagne *viticole* among cellar workers around Épernay.[68] Almost 150 men and women cellar workers at the Union champenois of Épernay, which had been targeted as a fraudulent producer by the "Livre noir," began a strike, which was quickly joined by workers from other champagne houses throughout the city in late March and early April 1911.[69] Although the workers did not directly link their wage demands or grievances with the problems of the vignerons, strikes were declared at firms that earlier police reports had indicated were particularly in need of protection from angry vignerons.[70] More important, according to the prefect, these cellar workers "were not specialists but simply workers recruited from among the vignerons without work. They are all devoted to the mass of *viticulteurs*." It was believed that these workers were passing along information to their fellow vignerons about outside wines being brought into cellars.[71] Moreover, representatives of Confédération générale du travail (CGT) trade union alliance—which had strong organizational support from local railway workers—held a meeting with local strike leaders on April 7, 1911.[72] Rumors circulated that more widespread actions were planned. That same day, police reported that they had uncovered a plan to dynamite the cellars of Moët et Chandon, Eugène Mercier, Gardet, Gauthier, and Union champenois.[73] Authorities were concerned to learn that "strikers ... declared that they are inclined to march hand in hand with the vignerons."[74]

News of the Senate vote to reconsider and possibly rescind delimitations was telegraphed to the Marne Fédération on April 10.[75] By the night of April 11 and the morning of April 12, the situation was critical around Épernay, and the prefect requested that troops be moved into villages (figs. 18 and 19). During the course of that night, troops were massed in Épernay and Reims. Another squadron was scheduled to leave for Aÿ, a city targeted as a center of fraudulent production in the "Livre noir."[76] Those in charge of the military forces were instructed to follow the "measures of prudence issued in the ministerial circular of October 18,

1907."[77] The government, it appeared, was trying to avoid a repeat of the confrontations in the lower Languedoc in 1907.

By mid-morning, events were already spinning out of control. Groups of vignerons had destroyed the cellars of the firms of Delouvin, Jacquot, Menudier, Lemaire, and Perrier in Damery. Three of these firms had been targeted by the writer of the "Livre noir." The cellars of Castellane in Dizy, where the previous day, protestors had written "Down with *fraudeurs*!" across the door, were also destroyed.[78] Aÿ, however, which had been so intimately linked in the "Livre noir" with the big fraudulent producers, became the center of the protest movement on April 12, 1911. Located between Reims and Épernay to the north of the Marne River, Aÿ was a key transport and communication center linking the two main areas of champagne production and villages to the south and west. More concerned with protecting Reims and Épernay, the prefect dispatched only a small contingent of troops to Aÿ the night of April 11.[79] Troops lined the boulevard du Nord, the main thoroughfare of the town, while vignerons and other locals milled in the streets. Captain Tavernot, whose responsibility it was to protect Aÿ, reported that nonvignerons among the demonstrators appeared to fully support the vignerons.[80]

Tensions escalated when troops intercepted a group of protestors marching to Aÿ from the neighboring town of Dizy. Vignerons struggled against the soldiers using wine stakes pulled from adjacent vineyards. Outfaced by the military, the vignerons dispersed into the vineyards, but later entered Aÿ by an unprotected route.[81] News of the confrontation on the road from Dizy reached Aÿ, and protestors hastily fashioned barricades on the boulevard du Nord. Hundreds of protestors gathered to sign a petition to the government denouncing the potential end of delimitations.[82]

Outraged to learn of the confrontation between troops and protestors marching to Aÿ from Dizy, Émile Michel-Lecacheur, one of the directors of the Fédération (and son-in-law of a former leader between 1904 and 1910), joined the protestors on the barricades.[83] Rumors circulated that women and children had been injured by anxious troops attempting to turn the crowd back toward Dizy. Michel-Lecacheur would later say that this was what had initially prompted him to join the protestors.[84] His presence galvanized the movement. With no protest from the crowd, Michel-Lecacheur ripped the red flag from its pole. His appearance on the barricades focused vigneron protest and marked the first active involvement of a Fédération leader in a mass demonstration since violence had broken out in January. Michel-Lecacheur, who was both a member of Paul Déroulède's Ligue des patriotes and an officer in the reserves, became their leader.

Michel-Lecacheur made a rousing speech, in which he declared himself willing to lead the protest, but not one flying the red flag. The crowd appeared to concur, and such symbols of the Left quickly disappeared. Michel-Lecacheur, whom later reports referred to as a *nationaliste enragé*, was joined by others who were "ready to kill or be killed for the cause." Michel-Lecacheur viewed the popular uprising in purely patriotic terms. This was an uprising to protect the emotional tie between Frenchmen and the soil, a sentiment expressed much more complexly by the prominent intellectual Maurice Barrès when he defined the French nation as a supreme spiritual value. The French Revolution was seen by Barrès and Michel-Lecacheur's role model, the poet Paul Déroulède, as an "inspiration of energy." Lecacheur and his vigneron followers saw themselves as the guardians of a revolutionary patriotism at stake in the current crisis. Indeed, Michel-Lecacheur would later describe his fellow vignerons as heroic soldiers prepared to die for the *patrie*. In their minds, *patrie* meant the champagne terroir.

Meanwhile, with blasts from *fusées paragrêles* and beating drums, thousands of other vignerons from throughout the region descended on villages. Municipal councils posted their resignations in a sign of solidarity with the rural population, declaring that it was "impossible for our population to continue their slave labor as they have done for twenty years."[85] Additional troops were dispatched to Aÿ and Épernay to protect champagne firms. But five barricades of overturned trucks, trees, and barrels blocked troop movements on some of the key routes in Aÿ.[86] Troops at first refused orders to force their way past the barricades. In a repeat of events in the Midi, a tense standoff ensued, with the crowd watching the hesitant troops.[87]

A group of nearly a hundred protestors, "among whom were thirty women," attacked the firm of Deutz & Geldermann in Aÿ, which the "Livre noir" called one of the key perpetrators of fraud, saying that it had close ties, not only with German producers of counterfeit champagne, but also with vineyards and production facilities in the Rhine region. Protestors were vicious. With rocks and bottles hailing down, troops sent to protect the firm were forced to retreat. By 2:30 that afternoon, there were reports of hand-to-hand combat in the rue Gambetta. Cut off from some of the principal streets in Aÿ, troops were unable to stop groups of protestors from entering the Gauthier cellars. Another group of protestors was intercepted attempting to set fire to the Pommery vine presses on the edge of town, where, according to the "Livre noir," grapes were processed for counterfeit champagne and honest vignerons were cheated.[88] Meanwhile, requests funneled in for protection of another firm targeted by the writer of the "Livre noir," that of the family of Deputy Lannes de Montebello in Mareuil, which was under siege, with troops already engaged in a bloody battle with an angry crowd. Once troops had been dispatched to Mareuil, however, the remaining squadrons in Aÿ were unable to hold back protestors, who set fire to the establishments of Ayala and Bissinger. With flames shooting out of the gutted buildings, protestors moved to the cellars of the firms of Van Cassel, Gallois, Ducoin, and Deutz & Geldermann—all previously targeted by the "Livre noir"—emptying thousands of gallons of wine in the streets and destroying equipment, bottles, and account books.[89] By 8:00 that night, fires from the Gallois building and other burning buildings along the rue Jeanson lit up the sky.[90] Reinforcements were sent, but they found the streets nearly impassable. "The entire length of the boulevard du Nord is strewn with debris of all sorts," one reporter noted, "broken and burnt furniture, papers and account books, materials from the cellars, barrels and crates of champagne . . . a safe ripped open and lodged in the middle of the ruins."[91] Among the debris was a dead horse, killed by a *fusée paragrêle*, and there were "numerous wounded" soldiers and protestors.[92]

Telegrams flowed into the prefect's temporary headquarters in Épernay, signaling the spread of the protest movement in smaller towns around Aÿ and Épernay.[93] While protestors were entering the cellars of Deutz & Geldermann in Aÿ, others, led by a group of women from the villages surrounding Moussy, were heading toward Épernay. Passing the town of Pierry, they entered the Rondeau establishment and within twenty minutes had broken machines and emptied cellars. A tense confrontation between agitated protestors and troops was ended with the intervention of the mayor of Pierry. The crowd of women and children scattered in the vineyards, but they reassembled several hours later in Épernay, which they entered by way of several unprotected streets.[94]

Under the control of General Abonneau, troops confronted protestors in the streets of Épernay. The angry crowd was headed in the direction of the maison Mercier. Troops were ordered to disperse the crowd. "We cannot convey the horrible sensation that we experienced at the scene," wrote one team of journalists, "the inhabitants of the rue Eugène-Mercier opened their doors to the vignerons and vigneronnes who were attempting to evade the trampling of horses and the sabers of troops."[95] Another journalist reported, "A woman, mouth open, cried out in anger and exasperation. A blow from a saber knocked her to the ground. Bloody, she raised herself up but was hit a second time; she fell."[96] Injured men, women, and children were reported to be scattered across the city. By 4:30 P.M., however, General Abonneau was able to report to the prefect that "the situation is clear in Épernay."[97]

Protestors limped home—bloody, beaten, exhausted. With the arrival of additional troops to occupy the vineyards, the protestors were clearly outnumbered. Violence subsided. Over the next week, there were a number of small protests and several tense confrontations between vignerons and négociants suspected of buying outside wines.[98] But, for the most part, the vineyards were overtaken by an uneasy calm. Marcelin Albert, the famous "savior" of the vignerons of the Midi, reportedly sent a telegram to the Fédération advising: "*Vignerons champenois,* one does not persuade with fire. . . . Your best weapon is prudence."[99] Reports continued to flow into the prefect's headquarters in Épernay: damage was extensive. Not only had the contents of champagne houses in Aÿ, Épernay, and surrounding villages been destroyed, but between 35,000 and 40,000 vines had been trampled or burnt in the vineyards between April 11 and April 14.[100] Although the names of property holders were not always listed in reports, the evidence suggests that it was mainly the vines of négociants and other large landowners that were attacked. The destroyed vines were almost all covered with the expensive tarps used to protect against frosts—tarps that Raoul Chandon had said were beyond the budget of the majority of vignerons.[101] Vignerons appeared to single out the property of large landowners and négociants for destruction. Those targeted figured prominently in the "Livre noir." Although there is no evidence that actions were planned or directed, evidence shows that only in a few rare cases did the crowd attack a house not targeted by the Comité des vignerons ruinés de la Marne. More important, perhaps, all reports indicate that protestors often bypassed the property—reportedly taking off their caps and lowering their protest flag in a show of respect—of négociants who were not deemed to be fraudulent producers.[102] The protest was spontaneous, but the violence was not random.

The repression that followed was less selective. Additional troops were sent to the vineyards, dividing the region into seven sectors to control the main roads and secure the safety of the firms. "In Champagne," a journalist wrote in *L'Humanité*, "there are more soldiers than vignerons."[103] Fear came with repression. There was no "arrogance" among the vignerons stopped for questioning by police. "They tremble," wrote Captain Diez, "and it is with difficulty that responses are obtained without hesitation to the most simple questions."[104] The authorities interrogated hundreds of protestors.[105] Witnesses were marched into a local theater to identify protestors by means of a confiscated newsreel.[106] A local journal would declare the action of police, investigators, and courts "excessively severe." "Without distinction," the writer continued, "the press of Reims raises its voice over the excessive severity of the arrests . . . and the sentences pronounced by the tribunal of Reims."[107]

Over the next several months, forty-six vignerons were convicted of crimes and two distraught protestors hanged themselves in their cells before they could be put on trial.[108] Most of the judiciary records are still sealed or were destroyed during the two world wars. Information gleaned from available sources, however, shows that participation in the protests of April 12 came from throughout the rural community. A review of the petition signed by protestors on April 12 in Aÿ, along with later arrests, indicates that a large portion of the crowd came from surrounding villages.[109] Of the thirty-three accused of taking part in the protests in Aÿ, there were twenty-four vignerons, one vigneronne, two *ouvriers-vignerons*, or hired vineyard workers, and seven local artisans, including those employed in cellars but who also owned vines.[110] There were similar breakdowns by profession for arrests and trials of those involved in protests in the towns of Pierry, Damery, and Vinay.[111] Although arrests were mainly of men between the ages of twenty and twenty-five, photos and film reveal that protestors were not confined to this age or gender group. As Jacqueline Saillet concluded in her analysis of the protests of 1911, the movement appeared to be "an affair of the *monde viticole.*"[112]

This viticultural world was the regional electoral base of the Radical party. Not surprisingly, Radicals who represented the Marne in Paris were quickly dispatched to Aÿ and Épernay. They immediately met with members of the Fédération to discuss the situation, and only afterward did they go on to a meeting with the prefect.[113] "All measures that will return calm and security among the unfortunate population," they declared, would be instituted quickly.[114] But with the heavy concentration of troops and the threat of more arrests, the protest movement was effectively stifled. By

June 1911, the Fédération once again appeared to lead the vigneron community.[115] But in many ways, the arrest of Michel-Lecacheur as a leader of the April protests in Aÿ generated the most support for the Fédération. Michel-Lecacheur's arrest gave the organization legitimacy with vignerons who identified with his nationalist call to protect champagne as the *patrie*. While the community awaited a decision from Paris on the final fate of delimitations, the release of Michel-Lecacheur became a focal point for the energy of diverse members of the rural community.[116]

With discussions of delimitations taking place during the months of May and June (the period of *travail fou*, or *travail à quatre bras*, when vignerons worked long hours) vignerons appeared to leave discussions of delimitations and amnesty for prisoners to the Fédération.[117] The ability of the Fédération to elicit promises from the Syndicat du commerce to undertake negotiations during the upcoming harvest undoubtedly added to vignerons' support of calls for calm.[118] After all their struggles, however, more radical members of the vigneron community were not completely silenced. Members of Comité des vignerons ruinés, the publishers of the "Livre noir," issued a statement to the minister of commerce in June 1911 declaring that their organization was "resolved to apply every means at their disposition for maintaining the rights acquired over twenty years of struggle and show to all of France that *les champenois marnais* are not sheep but men."[119] The compromise agreement reached in parliament to retain delimitations but create a Champagne *deuxième zone* in the Aube, however, produced no such collective protest. The government in Paris was clearly willing to protect regions, but still had difficulty achieving consensus on what defined a region within the modern nation-state.

No efforts were made to bridge the divide that developed within the community of négociants during the protests and violence of April. Négociants in the Syndicat du commerce attempted to distance themselves from the invective of the Syndicat de défense. "The Syndicat du commerce has struggled, without any other thought than maintaining solidarity with the vignerons, to obtain all measures relative to delimitations and the repression of fraud," the organization stated in a meeting of the Conseil général.[120] In a series of bitter exchanges between Bertrand de Mun, president of the Syndicat du commerce, and the leaders of the Syndicat de défense, Mun accused the editors of *La Champagne commerciale* of inciting violence and, through its rejection of delimitations, pushing vignerons into taking action. Leaders of the Syndicat de défense replied to these attacks by arguing that the smaller négociants were more in step with the cause of the vignerons. Declaring that "the grand Syndicat, composed of thirty négociants, has treated the hundreds of négociants [of the Syndicat de défense] as adversaries," the Syndicat de défense confirmed that it would continue to defend its interests and those of its members against the *grosses commerçants*.[121]

"The enemy is not the small proprietor [*petit patron vigneron*] who is, on the whole, only a laborer working for his own wages," Jules Moreau explained to a small gathering of vignerons in February 1912. Months after the first cellar was invaded in the "revolution in Champagne," there was still a search for the "dreaded enemy." "The enemy is Capital," he continued, "that is, the large wine houses [*grosses maisons*], which earn millions and construct châteaux while the proprietors die of hunger." The events of 1911 demonstrated the need for "a unity of laborers and *petits patrons vignerons*," whose interest was to cooperate in the struggle against "the oppression" of the négociants.[122] Moreau, who had just been acquitted after a highly publicized trial for participation in the events of early 1911, emerged to lead a new wave of vineyard protests. Abandoning the discussion of solidarity based on terroir—which could unite or divide capital and labor—Moreau attempted to focus attention on solidarity based on class. Although this movement was weakened by internal conflicts, efforts at organizing strikes of vignerons employed in the vineyards of some of the largest négociants were effective in channeling the anger and energy of 1911 into a form of protest that appeared to escalate until the outbreak of World War I.[123]

What labor activists such as Moreau were suggesting was a new conception of solidarity that reflected the shifting reality of the champagne community. The mévente reduced property ownership among vignerons. Growing numbers of them had lost control of the means of production. For those who remained on the land, appellations and delimitations created a monopoly and gave them increased power to negotiate with the large wine houses. Many vignerons were essentially little more than wage laborers, their "soul" going into the exploitation of other people's vines.[124] The labor syndicats would represent the interests of these Champagne vignerons based on this understanding of their interests, forming an eventual link to the larger French labor movement. "The final struggle in Champagne," which the vignerons of Damery had heralded in their protest song of January 1911, would be not against fraud, phylloxera, or even regional négociants, but against capital in general.

Marching as part of the proletariat, under the auspices of the CGT, the activists promised a new "community" that united French labor, urban and rural.[125] But prior to the 1920s, vigneron solidarity remained based on a connection with the rural world of viticulture and the community of "champagne." Regardless of the actual economic inequality among the vignerons or between vignerons and local shopkeepers, they ultimately perceived themselves as a comradeship based on a connection with the land and a collective vision of Champagne. Vignerons, even those who combined property ownership with wage labor, did not identify with those who were exclusively engaged in wage labor. Moreover, there was a continued solidarity between nonvignerons and vignerons in local villages—the "Champagne community"—as is clear from the arrest records of 1911.[126] Central to the construction of community was the notion of class, but this was based on a connection with "the *monde viticole*."[127]

Activists such as Moreau and CGT organizers in the post-1911 period were right in identifying how the connections between vignerons and the land were changing. Reports indicate that the prolonged hardship of the mévente was compelling a growing number of vignerons to sell all or most of their vines and seek employment as permanent day laborers. Some were finding employment on the large tracts of vineyards amassed by négociants who were consolidating and expanding their holdings during the mévente. Some even found themselves working parcels owned by négociants in return for "loyalty" (that is, guaranteed work) each harvest season. "Several grandes maisons," one report noted, "consider this as the most effective way of combating the turmoil within the *milieux vignerons*, turmoil that . . . could threaten the future of Champagne."[128]

While the growth in the ranks of wageworkers complicated the social structure of the vineyards, it also strengthened the bonds within the *milieux vignerons*. During the mévente, even family members of *petits patrons vignerons* occasionally found "themselves employed during part of the year by the large proprietors."[129] As growing numbers of vignerons found themselves predominantly or even exclusively engaged in wage labor, occupational distinctions were increasingly recognized between the salaried and the unsalaried. Landowning *propriétaires-vignerons*, or *petits patrons vignerons*, were distinguished by these titles from *ouvriers-vignerons*, or wage laborers. Yet peasants continued not only to list their occupational category as *vigneron* but also to show a shared vision of community that transcended these categories.

The strike actions of 1912–14, although appearing to mark a break with earlier protests, revealed an ongoing unity in the vignerons' vision of community. As one early strike action in Verzenay involving 430 vignerons demonstrates, strikes were directed against the eight largest landholders, all of whom were négociants whose wine houses were represented in the Syndicat du commerce. This action was not simply about wages, pitting labor against capital. When the 430 vignerons in Verzenay joined together to paralyze work in these vineyards, the eight large négociant landowners refused to negotiate, believing that their refusal would bring the strike to a rapid finish. It was apparent within days, however, that this strategy was ill conceived. Vignerons in surrounding communities—particularly the so-called *petits patrons vignerons*—offered to negotiate temporary work contracts for the bulk of the striking laborers, sharing the economic burden of the strike and allowing the strikers to hold out until their demands were met.[130] The strikes, the subprefect in Reims said, were less about wages than the result of a perpetual *mentalité fâcheuse* —a collective "bad attitude"—with respect to the production of counterfeit champagne by négociants.[131]

Regardless of property ownership or wage employment, the vignerons shared a common concern with questions of control over the vines, the wines, and the Champagne community. The vision of the Champagne community articulated by the Fédération and its supporters in 1911 continued to unite peasant proprietors, laborers, and local artisans. The Fédération leadership, with tips from laborers, proceeded to search for violations of the delimitation regulations.[132] And at the end of 1912, vignerons affiliated with the Fédération organized a peaceful show of solidarity with their fellow vignerons still being detained for participation in the protests of 1911.[133] Along with strike actions and organized meetings, vignerons occasionally sent a direct message to négociants believed to be defying the law on delimitations by sabotaging wines, quietly dumping gasoline into unguarded barrels. "We must not forget that we are here, in a region that has been shaken by an unprecedented economic crisis; that there remains malice and anger against the principal *maisons de Champagne*," the prefect of the Marne observed in 1913.[134]

For their part, the grandes maisons of the Syndicat du commerce remained suspicious of the vignerons.[135] While the vignerons appeared united under the Fédération, the négociants seemed, on the surface, to be permanently split between the Syndicat du commerce and the Syndicat de défense. But with the reorganization of the Syndicat de défense into the Association syndicale des négociants in 1913, the two groups of négociants began to find common ground. Both were, in essence, opposed to enforcement of any delimitation regulations or any oversight of sparkling wine production.[136] The Syndicat du commerce, unlike the Association syndicale des négociants, never openly challenged the validity of the delimitation legislation, but it continued to oppose the role of Fédération representatives in regulatory activities.

Reports indicate that the shift in focus away from the delimitation legislation to the validity of regulatory methods temporarily halted the antagonisms that emerged between the négociants of the Syndicat du commerce and those of the Syndicat de défense. For the négociants outside of the Syndicat du commerce, the complaints lodged by large négociants about regulation aided in undermining the delimitation legislation, seemingly supporting their arguments that delimitations hurt business as a whole.[137] Although the two groups were unable to create a united front to counter the strength of the *milieux vignerons*, their unity on the issue of regulation created a two-pronged assault on delimitations that kept the future of the legislation in jeopardy when the Marne once again became Europe's battlefield in 1914.

For the vignerons, however, regaining control over Champagne was the only way to end their misery. The question of vigneron control of Champagne remained embedded in a discourse that was less about class than it was about protecting the *patrie* and the rural community. Nationalism in the vineyards in 1911, just as in Lamarre's formulation almost three decades earlier, was true to a "romantic 'left-wing' nationalism of the mid-nineteenth century, in which both universalistic liberal and anti-individualist deterministic ideas coexist."[138] The nationalism of the vignerons based in terroir seemed organicist and racist, deriving from Lamarre's anti-German invective and isolating the négociants as outsiders, foreigners to the vineyards of Champagne. The outbreak of World War I and the French government's sequestering of the property of "German" négociants must have only served to reinforce the logic of this. Yet we see, too, in the vineyards a discourse about the universalist tradition rooted in the French Revolution and central to the dominant Radical party, fiercely supported by these same vignerons. This contradiction might suggest, as Zeev Sternhell has argued, that these seemingly disparate political traditions in France could coexist in the same system of thought. Sternhell argues that these different strands within the French national tradition were always linked.[139] Attempting to understand the vignerons who followed Michel-Lecacheur in 1911 in terms of the traditional conflict of Left and Right obscures this coexistence.

As for the government of the Third Republic, emerging from the bruising battles produced by Boulangist veterans and anti-Dreyfusard militants, there were no easy solutions to the impasse over regulating the wine industry. The concept that regional wine production had somehow to be protected, not through trade barriers or other more traditional forms of economic protectionism, but through protective barriers around terroir was, however, firmly established. The internal dissension between regions and the threat of violent uprisings for the *patrie* in the rural backbone of the Third Republic contributed to the sense that France was a stalemated society. Yet the stalemate came not over whether regional agricultural specialties should be a protected part of the national heritage but over the actual details of the protection. The power of the state to impose hegemonic definitions of "nation" and "community" on the rural world was undermined by both its own inherent weakness and the power of rural France in forcing the centralizing state to negotiate an ongoing relationship with the region.

The post-1911 strikes never resulted in mass protests on the scale of the anti-phylloxera syndicate movement of the 1880s or the delimitations protests of the early twentieth century.[140] Historians interested in the syndicalist and socialists movements that emerged in the region in the 1920s have tended to see some continuity between the pre-1911 protests and these later organizational developments.[141] Certainly, the experiences of the anti-phylloxera syndicats and the success of the Fédération in negotiating delimitations formed the foundations for the organizational efforts of syndicalists and socialists after 1920. But in many ways, these post-1911 strikes and later socialist organizations could be viewed as a break with the contention chronicled here. Strikes after 1912 and labor movements in the postwar period were centered on issues of wages and control of work routine largely absent from these earlier conflicts.

In these clashes and confrontations within Champagne, the imagining of community was central to determining the response to and outcome of the conflict. The comradeship between wage-earning and non-wage-earning vignerons had changed very little since the 1880s, when Lamarre had rallied vignerons against the phylloxera syndicate. While one might perceive wage-earning vignerons as a rural proletariat, the vignerons themselves—regardless of their status as wage earners—had a different vision, identifying their interests with the rural community, not a larger movement of labor. In their efforts to organize cellar workers and others who used the Bourse du travail d'Épernay (the labor exchange), CGT and other labor activists had little success in convincing rural laborers that their interests were linked with those of the urban proletariat.[142]

Labor activists such as Jules Moreau and socialists such as Jean Jaurès conceived of the rural community in Champagne as having two distinct elements: on the one hand, the ouvriers-vignerons, or wageworkers, and *petits patrons vignerons*, who had a common economic interest because of the négociants' control of wine production, and, on the other, the grosses maisons, which controlled sizable estates as well as wine production.[143] The struggle was seen as one between labor and capital, and activists exhorted the ouvriers-vignerons to challenge the négociants by creating workers' syndicats that would constitute the vanguard in the class struggle in Champagne and, eventually, unite with the urban proletariat in France as a whole.

The Fédération, which included small shopkeepers, rural artisans, and *petits patrons vignerons*, had long been criticized as an excessively moderate *syndicat mixte*.[144] When attacked for failing to unite with skilled hired wine cellar workers, the Fédération leadership responded by asking, "What common interests, vignerons, could you have with the *cavistes* and *tonneliers*, who are not vignerons themselves?"[145] Labor activists in Champagne countered this long-standing position of the Fédération by arguing that the *petits patrons vignerons* was "really only a laborer working for his own wages."[146] Although not directly attacking the Fédération, which played such an important role in the 1911 movement, the labor activists were proposing an alterative form of solidarity.

From the vantage point of 1914, the mass demonstrations and residual tensions of 1911 appeared to have a certain continuity with earlier struggles. The production of champagne between 1870 and 1914 had united labor (grape growers, rural artisans, and hired cellar workers and their families) and capital (large landowners and *négociants-manipulants*) who had a collective interest in champagne as a consumer product. Yet as the conflicts and contention of these years demonstrate, groups within the industry sometimes differed over the definition and boundaries of the community, membership in the collectivity, and control (as both arbiter and protector) of the economic and cultural heritage of Champagne.

Given the emergence of a labor movement out of the 1911 riots, many writers have struggled to make sense of the 1911 violence in terms of the development of the Left in France. Was the protest movement in Champagne an example of class struggle, a "revolution" in Champagne?[147] Or was it a "war on fraud" by producers frustrated by years of government inaction?[148] Events show that class antagonisms were present throughout the protests of 1911 and remained in evidence afterward. Given the production techniques in Champagne, the issue of fraud could and, at times, did set labor against capital. Symbols of the Left appeared throughout the protests, but as Michel-Lecacheur's removal of the red flag suggests, the "leftist" cause was seen regionally in nationalist terms. With peasants of the Marne divided against those of the Aube and négociants within the Marne sharply divided among themselves, it would be difficult to conclude that this was a struggle of labor against capital.

"Fraud" was defined as producing champagne with wines and grapes from outside of the region; at the heart of the 1911 conflict was the problem of defining regional boundaries. Within the modern nation-state, these battles over regional boundaries could appear archaic. Regional boundaries within France had, at least in theory, been done away with by the French Revolution. Departments and new rationalized administrative units appeared in place of regional classifications. If regional boundaries existed, they did so as an administrative necessity, established and controlled by the state. While regional boundaries were readily replaced by administrative units, however, it took longer to supplant regional identifications. Over the course of the nineteenth century, transport networks, education, military service, market expansion, and print media fostered national integration and forged new concepts of national identity. By the early twentieth century, regional boundaries—and with them regional identifications—had, presumably, been superseded by attachments to the new nation of France.

The violence of 1911 indicates, however, that the periphery was not breaking up in the face of national integration. There continued to be an active process in which regional groups on the periphery reinvented and redefined themselves and their place within the nation. The mass demonstrations of 1911 resulted from differences within the wine industry over how the boundaries of the community were conceived, as well as differences at the state level regarding the place of "regional personalities" within the modern nation-state. This indicates that the periphery was not collapsing in the face of the centralizing state, pushing for national integration. Indeed, the deadlock at the national level attests to the power of the periphery in demanding recognition of its place within the nation. In the end, the "cause" for which so many vignerons were "ready to kill and be killed" was membership in the collectivity, membership that would give them control (as both arbiter and protector) of the economic and cultural heritage of Champagne.

Questions of membership and community often arise in national histories. "Imagining the French community," as James Lehning has noted, "meant determining both who would be included in the nation and what form that inclusion would take."[149] During the "revolution in Champagne" of 1911, members of the rural population of Champagne asserted their power in "imagining the French community" and determining how they, as protectors of one of the glories of France, would be included in the French nation. If the task of the French nation was to integrate the separate *pays* that made up the rural world into the nation in a process "akin to colonization," as Eugen Weber has suggested, then the rural population laying claim to Champagne was involved in a full-blown "colonial" revolt in 1911.[150] The process of champagne becoming French would require that France itself undergo redefinition. In this case, revolt and resistance became the tools for forcing a redefinition that required acceptance of newly valorized regions. The regions would be reborn within the republic.

CHAPTER *Seven*

CONCLUSION

Champagne and Modern France

"In pursuing this protection of the appellation Champagne and the other vinicultural appellations we do not devote ourselves simply to platonic expressions destined to satisfy the vanity of our producers," wrote one legal expert and supporter of the new appellations legislation. "This is a national effort that must be earnestly carried out."[1] The sparkling beverages flowing from Champagne were more than simple wines. As Raymond Poincaré, president of the French republic during World War I, maintained, champagne was "a part of the French national fortune."[2] The champagne vignerons had shown themselves "ready to kill and be killed for the cause" that, for them, was purely patriotic. For those who took to the streets in 1911, defining the region, the specific areas or terroir within which a wine could be produced, was tantamount to protecting the *patrie*. By the eve of World War I, bordeaux, banyuls, and clairette de Die wines enjoyed similar protective legislation based on the principle of geographical delimitation. The region, now honored for its contribution of local agricultural specialties, was a protected part of the patrimony under the appellation d'origine contrôlée, or AOC, laws.

As if to seal this pact between nation and region symbolically, some of the most famous battle lines of the Great War crisscrossed the vineyards of Champagne. The land was infused with the blood of thousands of soldiers who killed and died for the *patrie*. The cause of the *patrie* was itself, in the minds of many of the most passionate of Déroulède's Ligueurs, like Champagne's Michel-Lecacheur, about the return of the soil of France's regions—the lost provinces of Alsace and Lorraine—to the nation. Two virulent, albeit very different, conflicts over regions within the nation were seemingly coming to a close.

During the war, Champagne vineyards were pulverized under the relentless assault of modern armed combat. The debate about whether or not to replant with American vines was over. Champagne was the last area of France to be infected with phylloxera, the last arena of debate over the controversial issue of whether American rootstock could make fine French wine. War destroyed a great part of the Champagne vineyards and with them the phylloxera aphids. Devoted growers, some living in the deep chalk caves that had served as champagne cellars, desperately attempted to care for vines growing in the midst of the war zone. When the shelling finally stopped, replanting was no longer an option but a necessity. In the 1920s and 1930s, few growers, even those who had the financial resources of the large wine houses behind them, wanted to take a chance on replanting with the vulnerable French rootstock, even if it meant loss of the esprit gaulois for their vin gaulois. Only 8,500 hectares of vineyards were replanted in the immediate aftermath of war, continuing the dramatic reduction in the area of production that had begun with the phylloxera. The former battlefields, cleared by American soldiers, were replanted with American rootstock grafted with French scions.

The war and the earlier Champagne riots did not, however, mark the end of the debate over fraud and adulteration. As the stalemate over the actual details of the legislation in the years before the war indicates, the quality of France's most famous wines was determined by more than simple geography, regional identifications, or carefully delimited terroirs, regardless of how much these were supposed to manifest France's spiritual values. Grape varieties, cultivation techniques, and poor winemaking practices could obscure the coveted essence of terroir. In the 1920s, the French wine industry struggled to reach an accord on a new, more detailed set of rules for wine appellations that would take into account, not simply geographical delimitations established prior to the war, but also vine varieties, cultivation techniques—such as pruning—and minimum alcohol strength. The Champagne region would also be part of these new accords, although the region remained mainly focused on issues of geography until the 1930s.

Although a complete study of champagne's history in the twentieth century remains to be written, France's most famous elixir undoubtedly continued to "oil the wheels of social life" and quench the thirst of privileged consumers after the war. Advertising became more sophisticated but in many ways continued to build on themes pioneered between 1870 and 1914—association with luxury, celebration, transport (extending from air travel to the increasingly popular automobile),[3] modernity, sports, and, always, France itself. Labels, posters, and other advertising media featured champagne's regional roots within France, and popular nineteenth-century brands continued to have widespread appeal for consumers. "Around the world," a 1935 Mercier advertisement proclaimed, "every 12 seconds a cork pops . . . from Champagne Mercier."[4] Sales figures over the course of the twentieth century did not necessarily keep pace with this claim of a cork every twelve seconds. Champagne, however, seemed to have captured the popular imagination of a mass consumer market well beyond the French hexagon, even in places where a cork might rarely be unleashed.

This book has focused on the process that transformed champagne and France's regional terroir into national possessions. The intrinsic belief that French wines deserved special status in the global marketplace came from both a popular shift in attitudes toward wine and its meaning and, in no small part, from the successful marketing of regional specialties as national goods. Wine appeared as something exceptionally French, to be protected against global market shifts. Elite wine and, in particular, formerly local specialties available to a limited clientele profited from improvements in transport, new processing techniques, and reduced production costs, which made it possible to market to ever broader groups of consumers both at home and abroad. Savvy marketers and taste professionals, both of whom touted these products as essential to the art of eating and drinking, made them a part of the culture of ingestion. As Pierre Bourdieu has argued, elite wine and French culture are so intertwined that it is today almost impossible to discuss the one without the other.[5]

Négociants joined with peasant growers to lobby the French government for domestic legislation. Their letters and petitions attempted to link the fate of French regional specialties, particularly luxury wines, with that of the nation. Pamphlets, letters to the editor, and press releases directed at a larger public stressed the ways in which wines like champagne were rooted in the very soil of France. By the early twentieth century, the land in France had an unshakable popular reputation, affirmed by writers, geographers, historians, and gastronomes as uniquely rich, fertile, and productive. "French soil enjoys the privilege of producing naturally and in abundance the best vegetables, the best fruits, the best wines in the world," Escoffier declared.[6] Wines, just like the privileged land of the French, were presented in lobbying efforts as the common property of the nation. Peasants growers went as far as demonstrating their willingness to "kill or be killed for the cause" of protecting wine's sacred place within France, rooted in one of the most sacred of all French possessions—the soil. Private companies and local peasant organizations could unite the national good to a common imagined past of France. This imagined past in which the regions were freed from the tyranny of the ancien régime linked the continued existence of the regions to a national narrative actively promoted by the state.

Much of the wine legislation that consumers now take for granted was a direct result of the contention and compromise outlined in this study. By 1911, mass demonstrations and violence had proven how far French vine growers were willing to go for the cause of protecting products of the soil. The resulting appellation d'origine laws had equally proven how willing the French public was to accept Champagne and other wine regions as worthy of a state effort to protect them. The granting of AOC protection to champagne and then to other wines and cheeses had proven how willing the state was to compromise with newly invigorated regions in the construction of the modern nation. The discourse of fine wine producers in Champagne had unleashed an intense debate at the turn of the century, which extended beyond wine production to include other foodstuffs linked to a delimited geographic area or terroir. This ruralist and protectionist discourse elevated the land and its products to a "central part of every Frenchman's legacy."[7] Indeed, Pierre Nora notes in the introduction to the monumental study of French memory, that "the 'land' [in France] would not exist without the ruralist and protectionist movement" of this period.[8]

The political tension at the turn of the century over the protection of wine delimitations and appellations resulted, in no small part, from the fact that it required creating undisputed boundary lines within the nation. Boundaries were a major structural component of the nation-states that emerged from the seventeenth to the nineteenth centuries. States were experienced in allocating, delimiting, and demarcating borderlands.

These boundaries became central to political nationalism, with its assumption of an opposition of nationalities, an "other" across the border. "The French Revolution," as Peter Sahlins notes, "gave the idea of territory a specifically national content, while the early nineteenth-century states politicized the boundary line as the point where national territorial sovereignty found expression."[9]

Internal boundaries were a different matter. In postrevolutionary France, boundaries were created as an administrative convenience meant to serve the needs of the centralizing nation-state. Even if new internal boundaries conformed to historical partitions, a sense of national identification was to transcend these borders, serving as a common denominator that defied prior regional, social, or political divisions. According to standard scholarship on France, this sense of national belonging integrated diverse and sometimes hostile groups during the early Third Republic (1870–1914). Universal military conscription, new transport networks, and compulsory education were the primary vehicles that created a national consciousness and undermined local loyalties and a sense of local and regional attachments.[10] Rural communities in particular were, according to this model, integrated through the dissemination of urban values and large-scale economic change. Local cultural attitudes, economic practices, and political values gave way; national issues and values became central to rural communities.[11]

The contention over delimitations and appellations based on terroir suggests that during the early Third Republic, the state could not simply impose values or boundaries on rural communities. The demand for recognized internal boundaries based on a local contribution to the nation demonstrates that rural communities could redefine themselves and their local identities within the nation-state. National identity, as Sahlins found in his study of the Pyrenees, could be built from a "local process of adopting and appropriating the nation without abandoning local interests, a local sense of place, or a local identity."[12] Rural communities, like that of Champagne, could oppose and use the state for their own ends. Commodities like champagne were, after all, local products grounded in regional economic relations. Yet by 1900, regional producers had positioned themselves as protectors of these commodities as part of the national patrimony, wielding enormous power in the struggle for protectionist laws and forcing the state to open up negotiations over internal boundaries.

Boundaries were critical for wine. The more expensive and exclusive the wine, the higher the stakes were in delimitations. Within months of passage of the Law of 1905, the champagne industry took the lead in lobbying the government for "complementary measures" to demarcate wine regions. After receiving hundreds of petitions from angry peasant vine cultivators and négociants, the minister of agriculture was charged with the establishment of a commission to examine the possibility of delimiting the vineyards. A public hearing on the issue was held in 1908 to determine and demarcate the boundaries of the Champagne region based on local use of the term, historic production techniques, and natural factors like soil and climate. That meeting, however, turned into a battle over the construction of the French past and the very meaning of the internal boundaries created by the French Revolution.

The process was so contentious because delimiting wine territories and defining appellations involved distinction and differentiation within the nation. Internal boundaries, much like their external counterparts, could create a sense of difference, a sense of "us" and "them." This perception of difference between nations, particularly along international boundaries, could serve the interest of the centralizing nation-state and build a sense of collective identity.[13] Within the nation, an oppositional model of identity could work against the nation-building process. This was widely recognized by those who carefully watched the debate over the delimitation of Champagne. Politicians, economists, and vignerons in excluded regions denounced the "carving" of France into separate, protected units. Appellations and the delimiting of wine regions, they argued, recognized claims of special status among Frenchmen. This had not only economic but also very real political consequences because of the status of champagne as a national good.

Supporters of the Third Republic, following the tradition of republican writers such as Jules Michelet, insisted that loyalty to nation came before loyalty to region. But what if the region or its commodities were believed to be the embodiment of that nation? What if regions and the products of their terroir were, as Vidal de la Blache and other French geographers instructed, an expression of a living entity that was France? Loyalty to the region and loyalty to the nation would no longer stand in stark opposition. Indeed, they would become one and the same. Private companies and local peasant organizations in the Marne were uniting their regional commodity to a common imagined past of France. This imagined past, in which the regions were freed from the tyranny of the ancien régime linked the continued existence of the regions to a national narrative actively promoted by the state. Champagne producers were inserting their distinctive voices in the construction of the nation of France; they were inserting their voices into an "imagined community."[14]

The champagne industry's ability to successfully mask what were essentially local interests as national concerns convinced government officials of the need to protect champagne and other fine wines as national patrimony. AOC laws granted champagne official French citizenship, with special rights and protections within the nation. But in the process of negotiation and compromise, the nation had also become Champagne—and, quickly thereafter, the regional terroirs of bordeaux, cognac, roquefort, brie, and a host of other agricultural specialties. The arguments put forth for champagne's "Frenchness" were repeated by other regional communities following the debates of the early 1900s. Through the highly publicized debates over appellations and delimitations, the champagne producers helped to construct a common French "memory" that gave the regions and regional agricultural products a national significance. The legal protections granted to Champagne, and later demanded by other regions, institutionalized the place of the regions within the nation. Indeed, by evoking soil, history, and tradition, the wine producers made what were essentially social and cultural constructs appear "natural" and, therefore, justly protected by a rational set of regulations. Products of the terroir, whether or not they were actually consumed in France, became an important aspect of the idea of "Frenchness." Protecting this part of the national heritage emerged as a central preoccupation of the French in the twentieth century.

The conflict over the definition of champagne was finally resolved in 1927 when the region that produced it was strictly redefined to include a little over 400 communes in the Marne and Aisne and a few communes in the Aube. Apart from the boundaries, most of the other conditions regarding grape varietals—limiting planting to pinot noir, pinot meunier, and chardonnay—and cultivation and manufacturing methods remained unchanged until the 1950s.

With the economic depression of the 1930s and new threats from declining markets, surplus production, and the spread of vine hybrids, national legislation began to move beyond geography and self-regulation with the establishment of the Institut national des appellations d'origine (INAO) and the promulgation of new AOC rules. The INAO was charged with administering, regulating, and granting appellations d'origine controlées for wines and spirits and, eventually, cheeses and other foodstuffs. The majority of France's luxury wines date their AOC status to this period around 1936 and 1937. These AOC designations could be quite detailed, and many experts found them to be a reliable guide to quality. They could also be extremely vague—as is the case with the general appellations of bordeaux and champagne—with enormous variations in quality, revealing little more than where the wine was produced. What links all of these French categorizations across the twentieth century, however, is the commitment to protect the products' "Frenchness," based in regional terroir.

The French system is the model for current European Union wine legislation. In 1990, the INAO reaffirmed the concept of geographical appellations for France (rejecting the popular American practice of identifying vine varieties on wine labels) and adopted a strategy of preserving the terroir by unceasingly striving to prevent the use of French place-names for non-French products. The commitment to geographic delimitations at the end of the century is articulated in terms that are strikingly similar to those used before 1914. One of the most fascinating examples emerged in 1998 when a small village of 650 residents in Switzerland became a battleground for the European Union. This "tempest in a bottle"—as the *New York Times* dubbed the dispute—revolved around the use of the appellation of "champagne" by villagers in Champagne, Switzerland.

The villagers of Swiss Champagne contended that it would simply not be possible for consumers to confuse their product—still white wine in a screw-top bottle, priced at around five dollars—with the bubbly French wine. The French champagne makers' association, the Syndicat de grandes marques, argued that this was not the primary issue. Enlisting the help of the INAO, they argued that "champagne" is not only a trademark of France, a position embraced in European Union regulations regarding commercial property, but also a part of a French collective identity. "It's our patrimony and our collective trademark," argued Daniel Lorson, a spokesman for the French producers. "If we don't defend it, in a few years the word 'champagne' won't mean anything."[15]

After years of arduous negotiations and volumes of carefully crafted accords, it defies credibility to think that Swiss and French intransigence was motivated solely by economic interests. Indeed, the rhetoric on both sides—in press releases, statements by government officials, and newspaper editorials—indicates the extent to which nationalism shapes economic policy and business strategies. Commercial property, in this case, "champagne" wines, are designated by both sides as national patrimony worthy of government protection. Masking what are essentially local business interests as national concerns infuses the debate with a national urgency. The subsequent protectionist legislation—in this case, accords for the protection of commercial property—then disguises the social and cultural construction of the commodity in "natural" attire. Commodities, much like the nation, take on a timeless, eternal quality that is legitimated by state policy and reinforced by business and industry in promotional and lobbying efforts.

Outside observers might note, not without a slight touch of cynicism, that Western businesses and industries have often wrapped themselves in various national flags to further their interests. Rarely, however, do we ask questions about the social and cultural consequences of adopting these tactics. Nationalist rhetoric on the part of business, in this case, the wine producers of France in the current dispute over the protection of commercial property in Europe, is part of a larger dialogue about national identity, which has shaped notions of "Frenchness" both within France and abroad. The case of wine legislation illustrates the extent to which nationalism was grounded in the social and economic realities of agriculture and industry. Current European Union legislation, by adopting French legislation from the early part of the twentieth century, is thus based upon a nationalist policy that was intended to bolster the economic power of French wine producers. Observers should thus not be surprised to find the postnationalist move toward European integration, and seemingly neutral laws on practical matters such as regulating commercial property, mired in nationalist conflict.

Daniel Lorson, the spokesman for the French producers in the recent dispute, was, in many ways, echoing statements made by French producers over a hundred years ago: "All is lost, save honor. But honor—reputation—it is this great name: *Champagne,* and it is this name, at this very moment, that they attempt to wrest from you," René Lamarre declared in 1890. "I cannot repeat it enough: with the way that [wine] lists are drawn up today, *within ten years people will no longer be acquainted with the name Champagne.*" Honor, reputation, patrimony—these were tied to the fate of the name "champagne." In the years leading up to the first laws concerning the protection of commercial agricultural property at the beginning of the twentieth century, the concerns of French producers were not very different from those of the contemporary wine lobby.

What has changed, however, is widespread acceptance of the assumption that producers and their regional agricultural products must be protected. No one in the current debate questions the position of wine as national patrimony, the national interest in protecting regional economic interests, or the reality of terroir and the claims for its delimitations. Indeed, in modern France, products with a claim to terroir are more popular than ever and the number of appellations granted to such diverse produce as honey, vinegar, and olive oils is on the rise. Names of villages, provinces, or *pays*—such as Champagne—continue to designate agricultural products that are considered local specialties. In many ways, these have become national "brands," protected much the way multinationals protect trademarks. While most consumers would not lose much sleep over the illegal use of corporate brand names, the usurping of "brands" for products connected to national place-names, to terroir, to products seemingly grounded in the soil of the nation can easily turn into a sustainable patriotic cause.

Popular notions about commodities linked to terroir and the associated AOC laws, much like notions of Frenchness themselves, have evolved and will continue to do so. Although the controversial laws governing delimitation of the champagne-producing region were revisited throughout the early twentieth century, the fundamental issue of whether champagne and regional agricultural products should be protected is not even questioned. Issues of how to best protect regional agricultural specialties continue to exist even today. While the social, economic, and political contexts have changed dramatically, the connection between the local—such as Champagne's wine, land, and people—and the collective imagining of the nation has not. In an era when the focus is on integration and a popular perception that the importance of nation-states is being eroded by economic globalization, these products of the terroir have only increased their centrality to a sense of collective identity. Debates over how to protect the products of the terroir may differ. But at least in France, the question of whether or not the nation has a responsibility to protect these products is beyond discussion. They have, one might argue, following the precedent of champagne, become French.

Appendix

Division of the Vineyards in the Marne

The champagne-producing region is divided into three primary zones.

THE ZONE DE LA MARNE

The Rivière Marne includes the vineyards around the towns of Cumières, Hautvillers, Dizy, Aÿ, Mareuil, and the entire right bank of the river to Avenay; continues west to the town of Château-Thierry.

The Côte d'Épernay includes the left bank of the Marne and the vineyards surrounding the towns of Épernay, Pierry, Moussy, and Vinay.

The Côte des Blancs or Côte d'Avize runs south of Épernay in the area surrounding the towns of Avize, Cramant, Mesnil-sur-Oger, and Oger. It includes the area around Vertus.

THE ZONE INTERMÉDIARE

The Zone intermédiare includes the towns of Bouzy and Ambonnay in the small area that separates the Zone de la Marne and the Montagne de Reims.

THE MONTAGNE DE REIMS

The Haute Montagne includes the area south of Reims and vineyards around Verzy, Verzenay, Sillery, Mailly-Champagne, Ludes, Chigny-les-Roses, and Rilly-la-Montagne.

The Basse Montagne includes the hills north of Reims.

CONVERSION FACTORS

    1 kilometer   0.621 miles
    1 hectare (ha.)   2.471 acres

1 liter	1.05 quarts
1 hectoliter	100 liters, or 26.4 U.S. gallons
la pièce	205 liters, or enough grapes to create approximately 273 bottles of champagne.

ABBREVIATIONS

ADM	Archives départementales de la Marne, Châlons-en-Champagne
AN	Archives nationales, Paris
APSGM	Archives privées du Syndicat des grandes marques de Champagne, Reims
U.S. consul, Reims	U.S. Department of State, despatches from U.S. consul, Reims, 1867–1906, microfilm T424, roll 1 (May 31, 1867–Dec. 31, 1886)

ONE ❦ INTRODUCTION

Epigraph: Karl Marx and Friedrich Engels, *Werke*, vol. 5 (1848; Berlin: Dietz, 1959), 462–63.

1. Pascal Ory, "Gastronomy," in Pierre Nora, *Realms of Memory: The Construction of the French Past*, ed. Lawrence D. Kritzman, trans. Arthur Goldhammer, vol. 2: *Traditions* (New York: Columbia University Press, 1997), 452.

2. L.-M. Lombard, *Le Cuisinier et le médecin* (Paris: L. Curmer, 1855).

3. Adolphe Brisson, preface to Armand Bourgeois, *Le Chansonnier du vin de champagne en 1890* (Châlons-sur-Marne: Martin frères, 1890).

4. See, e.g., the cover of *La Vie parisienne*, Jan. 2, 1909.

5. Elizabeth Olson, "A Tempest (but No Bubbles) in a Bottle of Swiss Champagne," *New York Times*, Dec. 9, 1998.

6. Hugh Johnson, foreword to James E. Wilson, *Terroir: The Role of Geology, Climate, and Culture in the Making of French Wines* (Berkeley: University of California Press, 1998), 4.

7. Michel Rachline, *La Férie du champagne: Rites et symbols* (Paris: Olivier Orban / Canard-Duchene, 1986), 11.

8. Roger Dion, *Histoire de la vigne et du vin en France des origines au XIXe siècle* (1959; Paris: Flammarion, 1977), 650.

9. Ibid.

10. Ernest Renan, *Qu'est-ce qu'une nation? Conférence faite en Sorbonne, le 11 mars 1882* (Paris: Calmann-Lévy, 1882), 26–29.

11. Dion, *Histoire*, 644.

12. Ibid., 34.

13. Ibid., 614.

14. Ibid., 646.

15. See, e.g., Rod Phillips's discussion of the "1999 Franco-Iranian Wine Incident" in *A Short History of Wine* (New York: Ecco, 2000), xiii.

16. Thomas Brennan, *Burgundy to Champagne: The Wine Trade in Early Modern France* (Baltimore: Johns Hopkins University Press, 1997), 273–72.

17. "Commerce des vins mousseux de Champagne" (official papers of the Chambre de commerce de Reims, n.d.).

18. Roland Barthes, "Rhetoric of the Image," in *Image, Music, Text* (New York: Hill & Wang, 1977), 32–51.

19. Raphaël Bonnedame, *Notice sur la maison Kunkelmann et Cie, Reims, Épernay, extraite des "Gloires de la Champagne"* (Épernay: R. Bonnedame, 1895). Published by *Le Vigneron champenois*, Sept. 1892.

TWO 🍇 CONSUMING THE NATION

1. Print reproduced in Henry Vizetelly, *A History of Champagne, with Notes on Other Sparkling Wines of France* (London: Henry Sotheran, 1882), 71.

2. Ibid., 109.

3. Eric Hobsbawm and Terence Ranger, eds., *The Invention of Tradition* (1983; New York: Cambridge University Press, 1992), 5.

4. Thomas Brennan, *Burgundy to Champagne: The Wine Trade in Early Modern France* (Baltimore: Johns Hopkins University Press, 1997), 273.

5. Roland Marchand, *Advertising the American Dream: Making Way for Modernity, 1920–1940* (Berkeley: University of California Press, 1985), xix.

6. Brennan, *Burgundy to Champagne*, 54–55.

7. Ibid., 64.

8. Ibid., 66.

9. Ibid., 44.

10. Ibid., 256–63.

11. Ibid., 249.

12. Daniel Pellus, *Femmes célèbres de Champagne* (Amiens: Martelle, 1992), 175.

13. Patrick de Gmeline, *Ruinart: La Plus Ancienne Maison de Champagne de 1729 à nos jours* (Paris: Stock, 1994), 63.

14. Brennan, *Burgundy to Champagne*, 272.

15. ADM, 18J, Moët.

16. Pellus, *Femmes célèbres de Champagne*, 178, quoting Victor Fiévet, *Madame veuve Clicquot (née Ponsardin), son histoire et celle de sa famille* (Paris: Dentu, 1865).

17. René Gandilhon, "Commerçants, vignerons et tonneliers champenois en Russie au cours du XIXe siècle," *Cahiers du monde russe et soviétique* 13 (1973): 502–13.

18. Pellus, *Femmes célèbres de Champagne*, 171.

19. Ibid., 178.

20. "Commerce des vins mousseux de Champagne" (cited ch. 1 n. 17 above).

21. Dion, *Histoire de la vigne et du vin*, 645.

22. Peter Gay, "On the Bourgeoisie: A Psychological Interpretation," in *Consciousness and Class Experience in Nineteenth-Century Europe*, ed. J. Merriman (New York: Holmes & Meier, 1979), 187–203.

23. U.S. consul, Reims, to assistant secretary of state, Jan. 15, 1869.
24. Robert Tomes, *The Champagne Country* (New York: Hurd & Houghton, 1867), 81.
25. Charles Tovey, *Champagne: Its History, Properties, and Manufactures, with Some Prefatory Remarks upon Wine and Wine Merchants* (London: Hotten, 1870), 48.
26. Adeline Daumard, *Les Bourgeois et la bourgeoisie en France depuis 1815* (Paris: Aubier, 1987), 54.
27. François Bonal, *Le Livre d'or de champagne* (Lausanne: Éditions du Grand-Pont, 1984), 63–66.
28. Daumard, *Les Bourgeois et la bourgeoisie en France depuis 1815*, 47.
29. William Reddy, "Need and Honor in Balzac's *Père Goriot:* Reflections on a Vision of Laissez-faire Society," in *The Culture of the Market: Historical Essays*, ed. Thomas L. Haskell and Richard F. Teichgraeber III (New York: Cambridge University Press, 1993), 335.
30. See Norbert Elias, *The Civilizing Process*, trans. Edmund Jephcott (1978; Cambridge, Mass.: Blackwell, 1994).
31. ADM, 16U194, labels nos. 16 and 44. For an example of the use of coats of arms in marketing, see the label for H. Moré and J. Savoye of Avize in ADM, 16U194, label no. 45.
32. ADM, 16U194, labels nos. 84 and 89.
33. Leo A. Loubère, *The Red and the White: A History of Wine in France and Italy in the Nineteenth Century* (Albany: State University of New York Press, 1978), 250–51.
34. ADM, 16U194, labels nos. 1169–72.
35. Ibid., labels nos. 57A and 63A, both deposited in 1869.
36. See Marchand, *Advertising the American Dream*, xvii and ch. 3; Jean-Paul Aron, *The Art of Eating in France: Manners and Menus in the Nineteenth Century* (New York: Harper & Row, 1975), 73.
37. Max Weber, *Max Weber on Charisma and Institution Building: Selected Papers*, ed. S. N. Eisenstadt (Chicago: University of Chicago Press, 1968).
38. Thomas Richards, *The Commodity Culture of Victorian England: Advertising and Spectacle, 1851–1914* (Stanford, Calif.: Stanford University Press, 1990), 84.
39. Clifford Geertz, "Centers, Kings, and Charisma: Reflections on the Symbolics of Power," in *Local Knowledge: Further Essays in Interpretive Anthropology* (New York: Basic Books, 1983), 122.
40. Daumard, *Les Bourgeois et la bourgeoisie en France depuis 1815*, 142.
41. Paul Butel, *Les Dynasties bordelaises de Colbert à Chaban* (Paris: Perrin, 1991), 245.
42. Tovey, *Champagne*, 101–2.
43. David Landes, "Business and the Business Man: A Social and Cultural Analysis," in *Modern France: Problems of the Third and Fourth Republics*, ed. E. M. Earle (Princeton, N.J.: Princeton University Press, 1951).
44. Tomes, *Champagne Country*, 95.
45. Reddy, "Need and Honor in Balzac's *Père Goriot*," 325–56.
46. Tomes, *Champagne Country*, 90.

47. Bonal, *Livre d'or*, 70–71.

48. Max Sutaine, *Essai sur l'histoire des vins de la Champagne* (Reims: L. Jacquet, 1845), 29–30.

49. See, e.g., ADM, 14U2133, "Modifications aux statuts de la Société Werlé," Sept. 27, 1913.

50. Georges Lallemand, "Édouard Werlé: Négociant en vins de champagne," *Champagne économique* 1 (1954): 3.

51. Ibid., 6.

52. David M. Gordon, *Merchants and Capitalists: Industrialization and Provincial Politics in Mid-Nineteenth-Century France* (University: University of Alabama Press, 1985), 52.

53. Ibid., 123.

54. ADM, 51M24, "Affaires diverses intéressant la sûreté générale: Surveillance des individus suspects et maisons étrangères, 1890–1917"; letter regarding the firm of Deutz and Geldermann, Dec. 8, 1914.

55. ADM, 197M23, "Statuts du Syndicat du commerce des vins de Champagne," art. 10.

56. Promotional literature and private correspondence with Y. Lombard, director general, Syndicat des grandes marques de champagne, Oct. 23, 1992.

57. Daumard, *Les Bourgeois et la bourgeoisie en France depuis 1815*, 52–54.

58. Patricia E. Prestwich, *Drink and the Politics of Social Reform: Antialcoholism in France since 1870* (Palo Alto, Calif.: Society for the Promotion of Science and Scholarship, 1988).

59. APSGM, Syndicat du commerce, Comptes rendus annuels, vol. 2, 1901–21.

60. ADM, Chp7065, Chambre de commerce de Reims, *Protestation de la Chambre de Commerce de Reims à l'égard des attaques dirigées contre le commerce des vins de Champagne (séance du mardi, 13 octobre 1908)* (Reims: Matot-Braine, 1908).

61. Octrois were city taxes, tolls, or duties.

62. APSGM, dossier 514.

63. APSGM, Syndicat du commerce, Comptes rendus annuels, vol. 2, 1901–21.

64. Roger Hodez, *La Protection des vins de Champagne par l'appellation d'origine* (Paris: Presses universitaires de France, 1923).

65. *Revue des deux mondes*, Oct. 1890; *Moniteur viticole*, Sept. 1873; *Vigneron champenois*, Apr. 6, 1898. See also reports in APSGM, dossier 7, "Affaire Kemp" (1909); dossier 9, "Olry-Roederer" (1908); dossier 25, "Cour d'Appel, Paris" (1895); dossier 37, "Jugement du Tribunal de commerce d'Angers, Werlé" (Aug. 20, 1869); dossiers 39 through 41, "Saumur"; dossier 42, "Cidre Avenal" (1901); dossier 57, "Lorraine-Champagne."

66. Sutaine, *Essai sur l'histoire des vins de la Champagne*, 145.

67. *Courrier de la Champagne*, Apr. 20, 1889.

68. Prestwich, *Drink and the Politics of Social Reform*, 304.

69. ADM, 16U194, labels nos. 1–71.

70. APSGM, dossier 692, "Resumé du travail des crus."

71. ADM, 16U194, labels nos. 17a–17c.
72. Ibid., labels nos. 71–81.
73. Ibid., labels nos. 440 ff.
74. Patrick Forbes, *Champagne: The Wine, the Land, and the People* (London: Gollancz, 1967), 115.
75. François Bonal, *Dom Pérignon: Vérité et légende* (Langres: D. Guéniot, 1995).
76. *Illustration*, Aug. 23, 1862.
77. See ADM, Chp6987, Syndicat du commerce des vins de Champagne, "Notice historique sur le vin de Champagne: Offerte aux visteurs de l'Exposition universelle de 1889" (Épernay, 1889).
78. Napoléon Legrand, *Le Vin de Champagne* (1896; 2d ed., rev., Reims: Matot-Braine, 1899).
79. Tovey, *Champagne*, 42.
80. Bonal, *Dom Pérignon*, 168.
81. Tomes, *Champagne Country*, 77.
82. Bonal, *Dom Pérignon*, 176.
83. *Petit Journal*, June 14, 1914, illustrated suppl.
84. Marchand, *Advertising the American Dream*, xviii–xix.
85. ADM, 16U194, labels nos. 2232, 683, and 636.
86. Rebecca L. Spang, *The Invention of the Restaurant: Paris and Modern Gastronomic Culture* (Cambridge, Mass.: Harvard University Press, 2000).
87. Aron, *Art of Eating in France*, 10–11.
88. ADM, 24J, Krug private archives.
89. Alfred Delvau, *Les Plaisirs de Paris: Guide pratique et illustré* (Paris: A. Fauré, 1867), 103–4.
90. Ibid., 103.
91. Prestwich, *Drink and the Politics of Social Reform*; James Roberts, *Drink, Temperance, and the Working Class in Nineteenth-Century Germany* (Boston: Allen & Unwin, 1984); Lilian Lewis Shiman, *Crusade against Drink in Victorian England* (New York: St. Martin's Press, 1988).
92. Michael R. Marrus, "Social Drinking in the Belle Époque," *Journal of Social History* 7 (1974): 120.
93. Brennan, "Towards the Cultural History of Alcohol in France," 78.
94. ADM, 16U195, labels nos. 1565, 2124, and 1616.
95. Charles Rearick, *Pleasures of the Belle Époque: Entertainment and Festivity in Turn-of-the-Century France* (New Haven, Conn.: Yale University Press, 1985), 201.
96. Erving Goffman, *Gender Advertisements* (1976; New York: Harper & Row, 1979).
97. ADM, 16U194, labels nos. 549, 2691, and 923.
98. ADM, 16U196, label no. 2804.
99. Bonal, *Livre d'or*, 146.
100. *Vie parisienne*, Apr. 21, 1881.
101. *New York Times*, Feb. 26, 1902; May 22, 1902.
102. Daumard, *Les Bourgeois et la bourgeoisie en France depuis 1815*, 28–29.

103. *Vie parisienne*, Aug. 20, 1905.
104. ADM, 16U195, labels nos. 1012, 1023, 1054, 1184, 1232, 1556, and 1702.
105. *Wine Trade Review*, Jan. 15, 1900.
106. ADM, 16U195, labels nos. 1615 and 1340.
107. ADM, 16U194, label no. 1220.
108. ADM, 16U195, label no. 2434.
109. Ibid., labels nos. 2107 a–c.
110. Reprinted in Bonal, *Livre d'or*, 173.
111. Roland Barthes, *Mythologies*, trans. Annette Lavers (New York: Hill & Wang, 1972), 58.
112. John R. Gillis, *Commemorations: The Politics of National Identity* (Princeton, N.J.: Princeton University Press, 1994), 10.
113. Ibid.
114. ADM, 16U194, label no. 522.
115. Ibid., label no. 671, Veuve de la Pleyne; label no. 729, Veuve Monnier et ses fils; label no. 734, Veuve Sillery; no. label 938, Veuve Fonteyne.
116. Ibid., label no. 692, Victoria; label no. 743, Royal Jubilee Champagne; label no. 808, Jubilee.
117. Ibid., no. label 957, Champagne Jeanne d'Arc.
118. Gillis, *Commemorations*, 10.
119. ADM, 16U194, label nos. 153–A, 591, and 592.
120. Briggs, *Wine for Sale*, 83.
121. *Revue des deux mondes*, Oct. 17, 1894.
122. Paul Leroy-Beaulieu in *Économiste française*, Apr. 22, 1911.
123. François Bonal, *Champagne Mumm: Un Champagne dans l'histoire* (Paris: Arthaud, 1987), 67–69.
124. See William Younger [William Mole], *Gods, Men and Wine* (London: Michael Joseph, 1966).
125. Gabriel de Gonet, *La Cusinière universelle: Nouveau livre de cuisine* (Paris: Le Bailly, 1883), 253.
126. Raphaël Bonnedame, "Quelque mots sur le vin de Champagne," *Vigneron champenois*, Sept. 1899.
127. Frédérique Crestin-Billet, *La Naissance d'une grande maison du Champagne: Eugène Mercier, ou l'audace d'un titan* (Paris: Calmann-Lévy, 1996), 162–63; Bonal, *Livre d'or*, 66–67. Photographs are preserved in the Bibliothèque municipale de Reims, Iconographique II, 7828 and 7906.
128. Loubère, *The Red and the White*, 250.
129. *Wine Trade Review*, Jan. 15, 1874.
130. Paul Leroy-Beaulieu, "Les Délimitations," *Économiste français*, Apr. 22, 1911.
131. Aaron Jeffrey Segal, "The Republic of Goods: Advertising and National Identity in France, 1875–1914 [i.e. 1918]" (Ph.D. diss., University of California, Los Angeles, 1995).
132. Semaine nationale du vin, *Compte-rendu de la Semaine du vin* (Paris: n.p., 1922), 378.

THREE 🍇 INDUSTRY MEETS *TERROIR*

1. ADM, Chp6874, René Lamarre, *La Révolution champenoise* (Paris: n.p., 1890), 43.
2. Ibid., 3.
3. Johnson, foreword to Wilson, *Terroir*.
4. *The Oxford Companion to Wine*, ed. Jancis Robertson (New York: Oxford University Press, 1994), 966.
5. Caroline Hannaway, "Environment and Miasmata," in *Companion Encyclopedia of the History of Medicine*, vol. 1, ed. W. F. Bynum and Roy Porter (London: Routledge, 1993), 299.
6. Jean-Anthelme Brillat-Savarin, *Physiologie du goût* (1825), trans. R. E. Anderson under the title *Gastronomy as a Fine Art; or, The Science of Good Living* (New York: Scribner & Welford, 1879), 3.
7. Allan Mitchell, "The Unsung Villain: Alcoholism and the Emergence of Public Welfare in France, 1870–1914," *Contemporary Drug Problems*, special issue, Fall 1986: 452.
8. Cited in Marrus, "Social Drinking in the Belle Époque," 120.
9. Jean-Yves Guiomar, "Vidal de la Blache's *Geography of France*," in Nora, *Realms of Memory*, 2: 188–89.
10. Menus reproduced in Auguste Escoffier, *Souvenirs inedits: 75 ans au service de l'art culinaire* (Marseille: Éditions Jeanne Laffitte, 1985).
11. APSGM, dossier 437.
12. Legrand, *Champagne*, 45.
13. Raoul Chandon de Brialles (a member of the Syndicat du commerce), "Le Vigneron champenois," *Revue de viticulture* 6, no. 158 (Dec. 26, 1896): 625–29.
14. Slava Liszek, *Champagne: Un siècle d'histoire sociale* (Montreuil: VO Éditions, 1995).
15. ADM, Chp6874, Lamarre, *Révolution champenoise*, 3.
16. One of the few studies on peasant holdings was undertaken by a representative in the French legislature in 1907. According to this report, there were 15,510 hectares of vines divided among 14,000 vignerons. See Herve Malherbe, "Socialisme et vignoble dans la Marne, 1890–1914" (master's thesis, Reims, 1974).
17. Raoul Chandon de Briailles, "Le Vigneron champenois," *Revue de viticulture* 8, no. 210 (Dec. 25, 1897): 685.
18. AN, F12 2489, Société nationale d'agriculture de France, *Enquête sur la situation de l'agriculture en France en 1879* (Paris: Imprimerie de Mme Ve Bouchard-Huzard, 1880), 5–6.
19. Robert Laurent, *Les Vignerons de la Côte d'Or au XIXe siècle*, 2 vols. (Paris: Les Belles Lettres, 1957–58).
20. Brennan, *From Burgundy to Champagne*, 15–18.
21. Ibid., 12.
22. Ibid., 13.
23. Georges Clause, "Le Vigneron champenois du XVIIIe au XXe siècle de

la pauvreté contestaire de 1789 à la révolution de 1911," *Études champenoises* 6 (1988): 22.

24. Raphaël Bonnedame, *Notice sur la Maison Moët et Chandon, Épernay* (Épernay: Vigneron champenois, 1892), 50.

25. André-H.-A. Jullien, *Topographie de tous les vignobles connus* (Paris: author, 1816).

26. Michel Hau, *La Croissance économique de la Champagne de 1810 à 1969* (Paris: Éditions Ophrys, 1976), 28.

27. Loubère, *The Red and The White*, 217.

28. ADM, 1J308, "Observations sur la culture de la vigne par M. Broupy, propriétaire et manoeuvre à Oger, près Avize, Marne."

29. ADM, 142M1, letter from subprefect, Reims, to prefect of the Marne, June 21, 1858.

30. Raoul Chandon de Brialles, "Le Vigneron champenois," *Revue de viticulture* 8, no. 210 (Dec. 25, 1897): 682.

31. Loubère, *The Red and the White*, 144.

32. G. Loche, "Délimitation de la Champagne viticole," *Revue de viticulture* 32, no. 835 (Dec. 16, 1909).

33. Sutaine, *Essai sur l'histoire des vins de la Champagne*, 98.

34. *Vigneron champenois*, Dec. 28, 1882.

35. ADM, 142M1, letter from the subprefect, Sainte-Ménéhould, June 22, 1858. There were only three hectares of vines remaining in this commune by 1900.

36. Armand Maizière, *Origine et développement du commerce du vin de Champagne* (Reims: L. Jacquet, 1896).

37. Jules Guyot, *Culture de la vigne et vinification* (Paris: Librairie agricole de la Maison rustique, 1860), 283.

38. Hau, *Croissance économique*, 27.

39. These figures are little more than estimates of production costs based on information found scattered in the records of the Syndicat du commerce and figures provided by Loubère, *The Red and the White*. Undoubtedly, costs were increased by phylloxera and other vine diseases, as well as by the increased use of fertilizers in the second half of the century. In fact, regional production costs jumped to between 2,800 and 3,500 francs per hectare by the 1890s. Contemporaries generally recognized that costs in Champagne were much higher than in other wine-producing areas.

40. ADM, 154M3, "Culture de la vigne," report to minister of agriculture, May 15, 1861.

41. ADM, 142M1, "Main-d'oeuvre agricole: Enquêtes, 1848–1927."

42. Hau, *Croissance économique*, 18; Louis Chevalier, *La Formation de la population parisienne au XIXe siècle* (Paris: Presses universitaires de France, 1950), 94.

43. Ministère de l'agriculture et du commerce, *Annuaire statistique de la France*, vol. 1 (Paris: Imprimerie nationale, 1878).

44. Hau, *Croissance économique*, 37.

45. ADM, 142M1, letter from M. Duguet to prefect of the Marne, June 10, 1858.

46. ADM, 142M1, letter from subprefect, Reims, June 21, 1858.

47. *Histoire des français XIXe–XXe siècles*, vol. 2: *La Société*, ed. Yves Lequin et al. (Paris: A. Colin, 1983); see also *Histoire de la France rurale*, vol. 3: *Apogée et crise de la civilisation paysanne de 1789 à 1914*, ed. Georges Duby and Armand Wallon (Paris: Seuil, 1975).

48. AN, F12 6405, Chambre de commerce de Reims, "Commerce des vins mousseux: Mouvement à partir de 1858."

49. For examples of these experiments, see Raoul Chandon de Briailles, "Le Vigneron champenois: Les Hommes," *Revue de viticulture* 9, no. 229 (May 7, 1898): 529–34.

50. Clause, "Vigneron champenois du XVIIIe au XXe siècle," 124.

51. Loubère, *The Red and The White*, 26–27.

52. Maizière, *Origine et développement du commerce du vin de Champagne*, 3.

53. APSGM, dossier 283.

54. *Vigneron champenois*, Dec. 22, 1873.

55. Comité central d'études et de vigilance contre le phylloxéra, *Enquête sur la vigne: Séance du 12 juillet 1883* (Épernay: Bonnedame, 1883).

56. Vizetelly, *History of Champagne*, 139.

57. Comité central d'études et de vigilance contre le phylloxéra, *Enquête sur la vigne*.

58. ADM, 147M31, commune of Aÿ, "État-matrice des cotisations à percevoir conformément à l'article 14 de l'acte constitutif."

59. ADM, 159M33, "Moyenne valeur d'un hectare de vigne depuis 10 ans, 1881–91."

60. ADM, E, suppl. 2581, Cumières.

61. Loubère, *The Red and The White*, 29.

62. ADM, 154M3, "Viticulture et viniculture: Renseignements sur la culture de la vigne et le product des récoltes, 1840–1934."

63. Maurice Crubellier and Charles Juillard, *Histoire de la Champagne* (1952; 2d ed., Paris: Presses universitaires de France, 1969), 386.

64. Vizetelly, *History of Champagne*, 208.

65. Lucien Lheureux, *Les Syndicats dans la viticulture champenoise* (Paris: Librairie générale de droit et de jurisprudence, 1905), 28–290.

66. See, e.g., ADM, E, *dépôt* 1506 F62, "Livrets d'ouvriers, 1892–1909."

67. *Revue de viticulture*, Aug. 1899, 295.

68. Loubère, *The Red and the White*, 217.

69. ADM, 18U6, minister of commerce to Conseil des prud'hommes concerning wage labor, June 20, 1890.

70. *Revue de viticulture* 7, no. 168, (Mar. 6, 1897): 249–58.

71. Raoul Chandon de Brialles, "Le Vigneron champenois," *Revue de viticulture* 6, no. 158 (Dec. 26, 1896): 629.

72. Lheureux, *Syndicats dans la viticulture champenoise*, 6–7.

73. The tradition of direct sales began among the most reputable vineyards in the mountain of Reims, but the practice was more widespread during the 1860s and 1870s, Vizetelly, *History of Champagne*, 152, observes. See also Bonal, *Livre d'or*, 89 n. 47.

74. Vizetelly, *History of Champagne*, 152.
75. APSGM, dossier 283. There were no available figures for the period before 1863.
76. Quoted in Bonal, *Livre d'or*, 91.
77. *La Vigne*, Sept. 20, 1877.
78. Syndicat du commerce des vins de Champagne, *La Culture de la vigne en Champagne* (Épernay: Bonnedame, 1889).
79. L. Néret, *Vertus: Glanes d'histoire* (Avize: n.p., 1916), 443.
80. *Vigneron champenois*, Dec. 22, 1873.
81. Quoted in Bonal, *Livre d'or*, 81.
82. Raoul Chandon de Brialles, "Le Vigneron champenois," *Revue de viticulture* 9, no. 230 (May 14, 1898): 561–65.
83. See ADM, 197M6, "Listes des syndicats, 1905–07."
84. ADM, J195, "Notice sur les récoltes vinicoles de Bouzy et d'Ambonnay de 1788 à 1874."
85. Raoul Chandon de Brialles, "Le Vigneron champenois," *Revue de viticulture* 9, no. 226 (Apr. 16, 1898): 437, fig. 62.
86. ADM, 18U6, Conseil des prud'hommes, June 20, 1890.
87. Jean-Alexandre Cavoleau, *Oenologie française, ou, Statistique de tous les vignobles et de toutes les boissons vineuses et spiritueuses de la France, suivie de considérations générales sur la culture de la vigne* (Paris: Madame Huzard, 1827).
88. Loubère, *The Red and The White*, 184.
89. Philippe Violart, *Le Calendrier du vigneron champenois* (Épernay: Bonnedame, 1877).
90. Bonal, *Livre d'or*, 629.
91. ADM, 83M30, "Police rurale: Détachements de gendarmes dans les communes à l'occasion des vendanges, 1881–90."
92. Violart, *Calendrier du vigneron champenois*.
93. Robert Smith Surtees, *Jorrocks' Jaunts and Jollities: The Hunting, Shooting, Racing, Driving, Sailing, Eccentric and Extravagant Exploits of That Renowned Sporting Citizen, Mr. John Jorrocks* (1838; London: G. Routledge, 1869).
94. C. Moreau-Bérillon, *Au pays du Champagne: Le Vignoble, le vin* (Reims: L. Michaud, 1925), 434.
95. Syndicat du commerce des vins de Champagne, *Notice historique sur le vin de Champagne* (Épernay: Bonnedame, 1889).
96. Raphaël Bonnedame, *Notice sur le vin de Champagne de la maison Pommery* (Épernay: Bonnedame, 1892).
97. U.S. consul, Reims, to assistant secretary of state on the cultivation of the vine within the department of the Marne, May 6, 1882.
98. *Vigneron champenois*, Oct. 8, 1878.
99. Raoul Chandon de Brialles, "Le Vigneron champenois," *Revue de viticulture* 8, no. 168 (Mar. 6, 1897): 249.
100. Ibid., 251–53.
101. P. Foureur, "La Saint Vincent fête des vignerons à Hautvillers," *Memoires*

de la Société d'agriculture, commerce, science, et art du département de la Marne 23 (1928–30): 109–30.

102. Malherbe, "Socialisme et vignoble dans la Marne," 10.

103. "La Politique," *Union républicaine*, Dec. 21, 1897.

104. B. Bourg-Broc, "Le Vocabulaire et thèmes politiques dans les affiches électorales marnais (1871–1914)," *Memoires de la Société d'agriculture, commerce, science, et art du département de la Marne* 85 (1970): 293 ff.

105. ADM, 7M58, 1888 elections.

106. ADM, E, *dépôt* 1506 F62, "Livrets d'ouvriers, 1892–1909."

107. Loubère, *The Red and the White*, 106.

108. Henri Jadart, *Reims—Guide: Visite aux monuments, aux maisons historiques, et aux principales curiosités de la ville* (Reims: F. Michaud, 1885).

109. Tovey, *Champagne*, 46.

110. Ibid., 61.

111. Ibid., 38.

112. For more on apprenticeships, see Maurice Hollande, *Chambre de commerce de Reims, 1801–1951: Un Siècle et demi au service de l'économie champenoise* (Reims: H. Debar, 1951), 44 ff., and Gandilhon, "Commerçants, vignerons, et tonneliers champenois en Russie." These patterns of apprenticeship were also common among the négociants of Bordeaux. See Butel, *Les Dynasties bordelaises de Colbert à Perrin*, 17–18, 116–17.

113. APSGM, Syndicat du commerce, Comptes rendus annuels, vol. 2, 1901–21.

114. See Bonnedame, *Notice sur la Maison Kunkelmann et Cie*.

115. ADM, 14U2133, "Werlé: Justice de paix, Modifications aux statuts de la Société, July 20, 1874."

116. Hollande, *Chambre de commerce de Reims, 1801–1951*. Madame Pommery's project met with disapproval from her associate Vasnier, who wrote, "Have you stopped the silly building at Crayères? It would be much better to stop than continue" (letter reprinted in Alain de Polignac, *Madame Pommery: Le Génie et le coeur* [Paris: Stock, 1994], 58).

117. ADM, Chp10189 and Chp10577.

118. *Vigneron champenois*, Oct. 20, 1899.

119. Vizetelly, *History of Champagne*, 139 and 229 ff.

120. U.S. consul, Reims, to assistant secretary of state, May 6, 1882 (cited n. 97 above).

121. Vizetelly, *History of Champagne*, 134.

122. Néret, *Vertus*, 443.

123. ADM, 197M4, "État des syndicats professionnels," Jan. 1, 1886.

124. ADM, 30M103, "Police—Politique: Royalistes," reports from the commissioner of police to prefect of the Marne, Sept. 8 and 29, 1879; ADM, 30M102, prefect to minister of the interior on legitimist banquets, Sept. 30, 1879, Oct. 22, 1879, and Dec. 5, 1879; ADM 30M103, "Police—Politique: Royalistes," subprefect, Reims, to prefect on meeting, Feb. 26, 1883; letter about royalist banquet, Feb. 28, 1883.

125. Tomes, *Champagne Country*, 95.
126. David Higgs, *Nobles in Nineteenth-Century France: The Practice of Inegalitarianism* (Baltimore: Johns Hopkins University Press, 1987), 111.
127. Tomes, *Champagne Country*, 78.
128. U.S. consul, Reims, to assistant secretary of state, Apr. 19, 1876.
129. ADM, 30M103, "Police—Politique: Royalistes," subprefect, Reims, to prefect on meeting, Feb. 26, 1883; letter about royalist banquet, Feb. 28, 1883.
130. ADM, 172M3 (107), "Situation industrielle et commerciale année 1878."
131. APSGM, letter to the minister of commerce from the négociants regarding consumption of champagne in the United States, Aug. 29, 1882, dossier 570; see also "Protestation du commerce des vins de Champagne," *Avenir de l'Est*, Sept. 23–24, 1882.
132. APSGM, Syndicat du commerce, Comptes rendus annuels, vol. 1, 1884–1900, "Rapport," Sept. 19, 1882; and ADM, Chp6880, Syndicat du commerce des vins de Champagne, "Rapport de Ch. Arnould, Secrétaire de la Chambre syndicale sur les travaux du syndicat" (Reims: Detraigne, 1884).
133. ADM, 197M23, Statuts du syndicat du commerce des vins de Champagne, art. 10.
134. Ibid., art. 1.
135. Sutaine, *Essai sur l'histoire des vins de la Champagne*, 145.
136. See sources cited in ch. 2 n. 65 above.
137. See, e.g., statement of syndicate in *Courier de la Champagne*, Apr. 20, 1889.
138. *Vigneron champenois*, Mar. 21, 1883.
139. APSGM, dossier 114.
140. ADM, J195 Bouzy; APSGM, dossier 283.
141. AN, F12 6405, "Commerce des vins mousseux" (cited n. 48 above).
142. ADM, Chp6874, Lamarre, *Révolution champenoise*, 10.
143. AN, F12 2489, Société nationale d'agriculture de France, *Enquête sur la situation de l'agriculture en France en 1883* (Paris: Imprimerie nationale, 1884).
144. One vigneron from Bouzy who sold his red wine for 600 francs in 1888 had by comparison reported getting nearly 5,500 francs in 1880, when he sold his grapes directly to négociants. See ADM, J195 Bouzy.
145. U.S. consul, Reims, to assistant secretary of state, Nov. 2, 1885.
146. *Vigneron champenois*, Oct. 8, 1888.
147. ADM, 154M17, "Extrait du procès-verbal des délibérations du Conseil municipal de Verzenay," Apr. 30, 1885.
148. ADM, 154M17, letter to the minister of agriculture from the prefect of the Marne, Mar. 1, 1895.
149. APSGM, dossier 283.
150. ADM, E, *dépôt* 1475 F31, Aÿ: "Statistique annuelle de 1880 à 1891."
151. ADM, Chp6874, Lamarre, *Révolution champenoise*, 6.
152. APSGM, dossier 283.
153. ADM, Chp6874, Lamarre, *Révolution champenoise*, 12.
154. Ibid.
155. Ibid, 24.

156. Ibid., 25.

157. APSGM, dossier 96, Lamarre, letter from René Lamarre to the Syndicat du commerce, Jan. 18, 1891; letter from Bollinger to the president of the Syndicat du commerce, Mar. 10, 1891; letter from the commissioner of police to Walbaum, Mar. 9, 1891.

158. APSGM, dossier 96, Lamarre, letter from Bollinger to the president of the Syndicat, Mar. 10, 1891.

159. Charles Gide quoted in Adrien Berget, "Producteurs et commerçants en Champagne," APSGM, "Pages libres."

160. Régis Daché, "Aux origines du mouvement ouvrier à Épernay, 1882–1914" (master's thesis, Reims, 1979), 17.

161. APSGM, dossier 96.

162. Malherbe, "Socialisme et vignoble dans la Marne," 26.

163. "Les Gens de Vincelles," *Figaro*, Aug. 20, 1891.

164. APSGM, dossier 96.

FOUR 🌹 RESISTANCE AND IDENTITY

1. Cartoon reprinted in "Le Phylloxéra en Champagne," *Union*, Aug. 17, 1992.

2. Karl Marx, *The Eighteenth Brumaire of Louis Bonaparte* (1852; New York: International Publishers, 1998), pt. 7.

3. I. Stevenson, "The Diffusion of Disaster: The Phylloxera Outbreak in the Department of the Hérault, 1862–1880," *Journal of Historical Geography* 6 (1980): 47–63.

4. Loubère, *The Red and the White*, 159.

5. R. Pouget, *Histoire de la lutte contre le phylloxéra de la vigne en France* (Paris: Institut national de la recherche agronomique, 1990).

6. Harvey Paul, *Science, Vine, and Wine in Modern France* (New York: Cambridge University Press, 1996), 19.

7. F. Cazalis, *Le Messager agricole* (Montepellier: Imprimerie de Gras, 1869), 237.

8. Marguerite Fitz-James, "La Vigne américaine," *Revue des deux mondes* 51 (1880): 889.

9. See, e.g., J. Barral writing in the *Journal d'agriculture pratique*, Sept. 20, 1868, 725.

10. Paul, *Science, Vine, and Wine in Modern France*, 24.

11. Escoffier, *Souvenirs inedits*, 191.

12. Lheureux, *Syndicats dans la viticulture champenois*, 37.

13. ADM, 1J203, Société d'agriculture, de commerce, des sciences et des arts de la Marne, "Phylloxéra," fols. 46, 48, 112, 138, 165, 180.

14. APSGM, Syndicat du commerce, Comptes rendus annuels, vol. 1, 1884–1900, "Rapport," May 13, 1886.

15. Bib. Mun. Châlons, Chp15939, Georges Vimont, *Rapport de la Commission internationale sur la question du phylloxéra en 1878: Conclusions à en tirer pour la culture de nos vignes champenoises* (Épernay: L. Doublat, 1879).

16. Malherbe, "Socialisme et vignoble dans la Marne." See also Maurice Agulhon, *The Republic in the Village: The People of the Var from the French Revolution to the Second Republic*, trans. Janet Lloyd (New York: Cambridge University Press; Paris: Éditions de la Maison des sciences de l'homme, 1982).

17. Bib. Mun. Châlons, Chp15938, Georges Vimont, *Album des syndicats champenois pour l'étude des principales maladies de la vigne et la recherche du phylloxéra (Comice agricole et viticole d'Épernay)* (Épernay: L. Doublat, 1881).

18. Abbé G. Dervin, *Six semaines en pays phylloxérés: Étude sur la défense et la reconstitution des vignobles français atteints du phylloxéra, suivie de la Champagne devant l'invasion phylloxérique* (Reims: Dubois-Poplimont, 1896), 282–84.

19. ADM, E, *dépôt* 2210F33, circular, Sept. 21, 1885.

20. Bib. Mun. Châlons, Chp15938, Georges Vimont, *Petit manuel & calendrier phylloxériques à l'usage des vignerons de Champagne (Comice agricole et viticole d'Épernay)* (Épernay: L. Doublat, 1880).

21. Dervin, *Six semaines en pays phylloxérés*.

22. ADM, 159M48/49, Mémoire justificatif présenté au nom des propriétaires vignerons à Monsieur le Ministre de l'Agriculture, Jan. 18, 1893.

23. *Vigneron champenois*, June 4, 1890.

24. Raoul Chandon de Briailles, *Recherches sur l'invasion du phylloxéra en Champagne* (Épernay: n.p., 1894).

25. APSGM, Syndicat du commerce, Comptes rendus annuels, vol. 1, 1884–1900, "Rapport 1889/90," May 21, 1890.

26. Ibid., "Rapport 1890–91," May 14, 1891.

27. Ibid.

28. ADM, 159M1, letter to prefect of the Marne, Aug. 19, 1890.

29. Bib. Mun. Châlons, Chp15808, Vimont, *Petit manuel & calendrier phylloxériques*.

30. Ministry of Agriculture, *Bulletin* 2 (1883): 457; *Bulletin* 9 (1890): 40–42.

31. See Napoléon Legrand, *Champagne*, 2d ed., rev. (Reims: Matot-Braine, 1899).

32. See Vizetelly, *History of Champagne*.

33. APSGM, Syndicat du commerce, Comptes rendus annuels, vol. 1, 1884–1900, "Rapport," May 7, 1885.

34. Hodez, *Protection des vins de Champagne*.

35. Cour d'Angers, Ann. 88.337 (11 Apr. 1887).

36. Laura Frader, *Peasants and Protest: Agricultural Workers, Politics, and Unions in the Aude, 1850–1914* (Berkeley: University of California Press, 1991).

37. APSGM, dossier 437.

38. *Vigneron champenois*, Apr. 6, 1898. For more on Russian sparkling wines, see APSGM, dossier 406, letters from the consul of France in Odessa and Eugène Gibert (agent of Moët et Chandon) in St. Petersburg, 1884.

39. *Revue des deux mondes*, Oct. 12, 1890.

40. ADM, E, suppl. 2581, Cumières, "Renseignements fournis à M. Bin par la Syndicat des vignerons de Cumières, 8 novembre 1904."

41. APSGM, dossier 57, Lorraine champagne made in Koblenz.

42. APSGM, Syndicat du commerce, Comptes rendus annuels, vol. 1, 1884–1900, "Rapport," May 14, 1891.

43. Georges Dumas, "L'Enseignement de la viticulture dans la Marne de 1890 à 1930," *Études champenoises: Vignerons et vins de Champagne et d'ailleurs XVIIe–XXe siècle* (Reims: Presses universitaires de Reims, 1988), 37–45.

44. The prefect of the Marne summarized these efforts to create the syndicat and acknowledged the enormous difficulty in obtaining signatures: "Réaliser cette oeuvre était difficile, dit M. Le Préfet, c'était presque tenter l'impossible, on le tenta: le Professeur d'agriculture secondé par l'administration se mit à l'oeuvre et à la suite de sérieux efforts parvint à réuni le nombre d'adhésions exigées par la loi." ADM, 159M1, "Rapport, Assemblée générale des propriétares de vignes," July 11, 1891.

45. Jean A. Barral, *La Lutte contre le phylloxéra* (Paris, n.p., 1883), 67–80.

46. ADM, 159M1, letter to mayors from the prefect of the Marne regarding syndicate, Sept. 27, 1890.

47. ADM, 158M48, letter to the minister of agriculture from the propriétaires-vignerons of the Marne, Feb. 18, 1893.

48. ADM, E, *dépôt* 2210F33, Archive communal, Dormans, circular, Sept. 27, 1890.

49. ADM, 159M1, letter to the prefect of the Marne from Comte Alfred Werlé, Armand Gustave Loche, Gaston Chandon de Briailles, and Henri Gallice regarding creation of syndicate, Aug. 16, 1890.

50. *Vigneron champenois*, Sept. 24, 1890.

51. ADM, 159M43, subprefect, Épernay, on hostility to syndicate at Vincelles.

52. ADM, E, *dépôt* 2210F33, letter from the mayor of Dormans to the prefect of the Marne, Oct. 8, 1890.

53. ADM, 159M43, subprefect, Épernay, to prefect of the Marne on local conferences with the vignerons, Oct. 10, 1890.

54. ADM, Chp6874, Lamarre, *Révolution champenoise*, 9.

55. Bib. Mun. Châlons, Chp15808, Vimont, *Petit manuel & calendrier*, 10.

56. "Les Gens de Vincelles," *Figaro*, Aug. 20, 1891.

57. "Le Sulfure de carbone," *Révolution champenoise*, Mar. 8, 1892; "Le Phylloxéra en Champagne," Ibid., Sept. 7, 1892.

58. "Le Phylloxéra," *Révolution champenoise*, Mar. 15, 1891.

59. ADM, 159M1, letter to mayors from the prefect of the Marne regarding property division in the Marne, June 5, 1891.

60. ADM, 159M1, letter from the prefect of the Marne to the minister of agriculture, June 27, 1891.

61. *Révolution champenoise*, July 21, 1891.

62. ADM, 159M1, letter to the minister of agriculture from the prefect of the Marne regarding the General Assembly, June 27, 1891. The number of vignerons is given in ADM, 159M48/49, "Mémoire justificatif," Jan. 18, 1893.

63. See, e.g., ADM, 159M48/49, "Mémoire justificatif," Jan. 18, 1893.

64. Ibid.

65. ADM, 159M1/2, letter from the prefect of the Marne to the minister of agriculture, June 27, 1891.

66. ADM, 159M1/2, letter from the prefect of the Marne, Aug. 16, 1890; ADM, 159M48/49, "Mémoire justificatif," Jan. 18, 1893; ADM, Chp6874, Lamarre, *Révolution champenoise*, July 21, 1891; ADM, 159M1/2, letter from the prefect of the Marne to the minister of agriculture, June 27, 1891.

67. *Révolution champenoise*, July 7, 1891.

68. Ibid., Aug. 25, 1891.

69. Bourg-Broc, "Vocabulaire et thèmes politiques," 287–327.

70. *Vigneron champenois*, Feb. 4, 1891.

71. APSGM, Letter from the president of the Syndicat to members, n.d., dossier 96; letter from Bonnedame to the Syndicat, Jan. 19, 1891, dossier 96; letter from Bollinger to the president of the Syndicat regarding police report on meeting in Aÿ, Mar. 10, 1891, dossier 96; letter to Walbaum regarding meetings of Lamarre, Mar. 9, 1891, dossier 96.

72. *Révolution champenoise*, Sept. 29, 1891; *Révolution champenoise*, Jan. 5, 1892.

73. ADM, 159M48/49, "Mémoire justificatif," Jan. 18, 1893.

74. ADM, 159M15, extract from *Vigneron champenois*, July 28, 1891.

75. *Révolution champenoise*, "Triomphe de la liste de protestation," Aug. 18, 1891.

76. Christian Lagauch, "L'Organisation syndicale des vignobles marnais de 1884 à 1946" (master's thesis, Reims, 1970), 31.

77. ADM, 159M15, the prefect of the Marne to the minister of agriculture regarding the nomination of the steering committee, Sept. 9, 1891.

78. ADM, 159M18, "Procès-verbal de la séance du comité directeur," Jan. 30, 1892.

79. Figures from "Commerce des vins mousseux de Champagne" (cited ch. 1 n. 17 above).

80. "Le Cas de M. Girard," *Figaro*, Mar. 20, 1889.

81. *Révolution champenoise*, Nov. 13, 1891.

82. ADM, 159M18, "Procès-verbal de la séance du comité directeur," Jan. 30, 1892.

83. ADM, 159M16, "Listes des syndics désignés par la Chambre de commerce à Reims," n.d.

84. ADM, 159M48/49, "Mémoire justificatif," Jan. 18, 1893.

85. Ibid.

86. ADM, 159M18, "Procès-verbal," Nov. 7, 1892.

87. Lagauch, "Organisation syndicale," 44.

88. ADM, 159M21, letter from the mayor of Venteuil to the director of the Syndicat contre le phylloxéra, Oct. 23, 1892.

89. Edmond Doutté, *Le Vignoble bordelais et l'avenir du vignoble champenois devant l'invasion phylloxérique* (Châlons, n.d.), 14–15.

90. See Ibid., 4–5, 14–15.

91. Dervin, *Six semaines en pays phylloxérés*, 291.

92. See, e.g., ADM, 159M49, meeting of Syndicat, Jan. 10, 1892, for the January incident. See also *Journal de la Marne*, Nov. 9, 1892, for a similar incident later that same year.

93. ADM, 159M21, letter from the mayor of Venteuil to the director of the Syndicat contre le phylloxéra, Oct. 23, 1892.

94. ADM, 159M22, letter from the prefect to the minister of agriculture regarding the resignation of the vigneron representatives, Dec. 29, 1892.

95. ADM, 159M42, letter to the prefect from E. Cochois, Aug. 8, 1891.

96. Ernest Mondet, "Communes de Tréloup et Passy-sur-Marne," *Révolution champenoise*, Oct. 6, 1891.

97. ADM, 159M42, petition to the members of the steering committee from the vignerons of Damery, Venteuil, Cumières, Fleury, Boursault, and Vauciennes refusing payment, 1892; letter from the mayor of Baslieux-sur-Châtillon to the prefect regarding contributions to the Syndicat contre le phylloxéra, Sept. 21, 1892; letter from the vignerons of Oeuilly to the prefect regarding contributions, Sept. 21, 1892.

98. ADM, 159M42, letter from mayor of Troissy to the director of the Syndicat contre le phylloxéra, Sept. 17, 1892; letter from subprefect, Épernay, to prefect on opposition in the vineyards, Oct. 6, 1892.

99. ADM, 159M48/49, "Mémoire justificatif," Jan. 18, 1893; ADM, 159M43, Commissariat spécial, report no. 16, Feb. 13, 1894.

100. "Incidents de Damery," *Révolution champenoise*, Oct. 13, 1892.

101. ADM, 159M49, "Extrait des registres du Conseil de préfet," June 19, 1893.

102. "Une Interview avec M. Werlé," *Indépendant rémois*, Nov. 8, 1892.

103. APSGM, Syndicat du commerce, Comptes rendus annuels, vol. 1, 1884–1900, May 5, 1892.

104. Ibid., May 4, 1893, and Dervin, *Six semaines en pays phylloxérés*, 323–24.

105. *Vigneron champenois*, Aug. 19, 1890.

106. ADM, 159M43, *Union républicaine*, Dec. 28, 1893.

107. ADM, 159M43, "Registre de la commune de Mardeuil," May 15, 1892.

108. Ibid., June 1, 1892.

109. Doutté, *Vignoble bordelais et l'avenir du vignoble champenois*, 14–15.

110. ADM, 159M43, petition to the prefect of the Marne from the vignerons of Vandières (arrondissement of Reims), Jan. 24, 1894.

111. ADM, X E2210F33, letter to the mayor of Épernay from the vignerons of the commune of Épernay, Mar. 30, 1893.

112. ADM, 159M48/49, "Mémoire justificatif," Jan. 18, 1893.

113. ADM, 159M42, letter from the mayor of Reuil to the prefect of the Marne regarding partisans of treatment methods.

114. ADM, 159M43, letter from the vignerons of the commune of Vandières to the prefect of the Marne, Jan. 23, 1894.

115. ADM, 159M43, "Contestations de la légalité du syndicat."

116. ADM, 159M43, Commissariat spécial, report no. 16, Feb. 14, 1894.

117. Ibid.

118. "Une Émeute en Champagne," *Le Temps*, Feb. 21, 1894; "Troubles à Vertus," *Franc Parleur*, Feb. 21, 1894; "À Vertus," *Courrier de la Champagne*, Feb. 20, 1894; "L'Agitation en Champagne," *Revue de viticulture*, Feb. 24, 1894.

119. ADM, 159M43, Commissariat spécial, report no. 31, Feb. 20, 1894.

120. ADM, 159M43, Commissariat spécial, report no. 40, Feb. 23, 1894; no. 47, Feb. 25, 1894; no. 64, Mar. 1, 1894.

121. ADM, 159M42, letters, 1895–96; and "L'Agitation en Champagne," *Revue de viticulture*, Feb. 24, 1894.

122. *Indépendant rémois*, Mar. 10, 1894; *Éclaireur de l'Est*, Mar. 10, 1894.

123. ADM, 159M20, report of the steering committee to subprefect of Épernay, July 15, 1895.

124. ADM, 159M48, convocation of the General Assembly, June 17, 1896.

125. Dervin, *Six semaines en pays phylloxérés*, 296–97.

126. Hau, *Croissance économique*, 51.

127. ADM, 159M33, letter from the propriétaires-vignerons of Broyes to the Syndicat contre le phylloxéra, Apr. 27, 1891.

128. Leo A. Loubère et al., eds., *The Vine Remembers: French Vignerons Recall Their Past* (Albany: State University of New York Press, 1985), 121.

129. ADM, 155M5, response of subprefect, Reims, to request of prefect of the Marne, Oct. 8, 1908.

130. ADM, E, suppl. 2574, Aug. 20, 1900.

131. Ibid., Aug. 27, 1909.

132. Frédérique Crestin-Billet, *Veuve Clicquot: La Grande Dame de la Champagne* (Paris: Glerat, 1992), 116–19.

133. ADM, 155M5, "Répartition de la propriété viticole de Reims," Oct. 8, 1908.

134. ADM, 155M10, "Enquête sur la crise viticole," 1901–2; responses from the Syndicat agricole et viticole du canton de Vertus.

135. ADM, 197M7, "Syndicats professionnels."

136. ADM, 1504F60, "Association viticole," Mar. 15, 1899.

137. See ADM, 197M7, viticultural syndicates created between 1884 and 1914.

138. Émile Lessard, *La Champagne: Agronomie, économie* (Reims, 1981), 32.

139. APSGM, Syndicat du commerce, Comptes rendus annuels, vol. 1, 1884–1900, May 5, 1898.

140. APSGM, dossier 657, "Subventions reçues par les Syndicats Antiphylloxériques"; letter to Association Viticole regarding aid to vignerons, 1908; letter from Mun to Giesler, Jan. 18, 1915.

141. ADM, 30M42–56; syndicate records in ADM, 140M2–3; and APSGM dossier 96, Lamarre.

142. *Journal officiel*, Mar. 26, 1892.

143. ADM, 7M60, election of 1898; *Revue de la Marne*, May 11, 1898.

FIVE 🟊 BOUNDARIES

1. Bonnedame, "Quelque mots sur le vin de Champagne."
2. For more on the ineffectiveness of early fraud legislation, see Loubère, *The*

Red and the White, ch. 12, and Charles K. Warner, *The Winegrowers of France and the Government since 1875* (New York: Columbia University Press, 1960), ch. 4.

3. Warner, *Winegrowers of France*.

4. For more on modern regulations of the industry, see Bonal, *Livre d'or*, ch. 13.

5. *Révolution champenoise*, Dec. 5, 1891.

6. Leo A. Loubère, *The Wine Revolution in France: The Twentieth Century* (Princeton: Princeton University Press, 1990), 114.

7. APSGM, Syndicat du commerce, Comptes rendus annuels, vol. 2, 1901–21.

8. APSGM, dossier 593, "Champagne contre Saumur."

9. APSGM, dossier 38, brief from the Cour d'Angers, July 19, 1887.

10. Paul Labarthe, *Dictionnaire populaire de médecine usuelle* (Paris: Marpon & Flammarion, 1889); excerpt in APSGM, dossier 124.

11. Ch. Tallavignes, "Les Vins en Champagne," *Revue de viticulture* 534 (Mar. 1, 1904): 500–503; Raymond de la Morinais, "Les Vins mousseux en Allemagne," ibid. 536 (Apr. 7, 1904); correspondence in APSGM, dossier 514.

12. *Révolution champenoise*, Nov. 13, 1891.

13. Ibid., Nov. 6, 1891.

14. ADM, E, *dépôt* 1475F31, "Distribution des plants americains" (memo from prefect), Nov. 20, 1900.

15. "Paris Day by Day," *Daily Telegraph*, Aug. 28, 1903; "A Poor Vintage," *Daily Telegraph*, Sept. 3, 1903; "Few Wines Are Pure," *Detroit New Tribune*, Jan. 17, 1904; editorial, *Chicago Chronicle*, Aug. 14, 1905; "The Truth about Champagne," *Times* (London), Nov. 19, 1908.

16. See Warner, *Winegrowers of France*, and Eugen Weber, *Peasants into Frenchmen: The Modernization of Rural France, 1870–1914* (Stanford: Stanford University Press, 1976).

17. APSGM, Syndicat du commerce, Comptes rendus annuels, vol. 2, 1901–21, meeting of May 8, 1903.

18. ADM, "Enquête sur la crise viticole," 1901–2; responses from the Syndicat agricole et viticole du canton de Vertus.

19. Marquis de Polignac, "On ne vendange pas," *Journal de l'alimentation*, Oct. 10, 1910.

20. ADM, 172M3, Chambre de commerce de Reims, document 30, commercial and industrial report, 1902 (July, 18, 1902).

21. See, e.g., discussion by Paul Leroy-Beaulieu in *Économiste française*, Apr. 22, 1891.

22. ADM, Chp10203; A. Perrin, *Étude sur l'oeuvre des syndicats contre la fraude des vins de Champagne* (Épernay: Imprimerie sparnacienne, 1907).

23. ADM, 155M1, petition from the commune of Hautvilliers, Aug. 16, 1903.

24. Ibid., petition from the commune of Verzenay, May 2, 1902.

25. ADM, 172M2, Chambre de commerce de Reims, document 30, Commercial and industrial situation, 1902 (July 18, 1902).

26. ADM, 155M1, extrait du procès-verbal des délibérations du Conseil général, Apr. 20, 1903.

27. ADM, 172M3, Chambre de commerce de Reims, document 28, commercial and industrial situation, 1902 (Jan. 22, 1902).

28. ADM, 172M3, Chambre de commerce de Reims, document 26, commercial and industrial situation, 1903 (July 7, 1903).

29. Yearly reports found in the private archives of the Chambre de commerce et d'industrie de Reims et d'Épernay and the semester reports of the Chambre de commerce du Reims found in ADM, 172M3.

30. ADM, Chp10203; Perrin, *Étude sur l'oeuvre des syndicats*.

31. Ibid.

32. ADM, E, *dépôt* 1475F31.

33. "Syndicat des vignerons," *Éclaireur de l'Est*, Jan. 30, 1904; ADM, 1504 M60, "Fédération des syndicats viticoles de la Champagne, 1904."

34. ADM, 197M7, list of viticultural syndicates between 1884 and 1914.

35. ADM, 197M6, letter from subprefect, Épernay, to prefect of the Marne, Jan. 28, 1905; report of subprefect, Reims, 1907.

36. ADM, 172M3, Chambre de commerce de Reims, document 26, commercial and industrial situation, 1903 (July 7, 1903).

37. *Observateur de l'Est*, Feb. 6, 1904.

38. "Appel aux vignerons," *Independant rémois*, Feb. 23, 1905.

39. Marx, *Eighteenth Brumaire of Louis Bonaparte*, pt. 7, 128.

40. "Vues politiques," *Revue de Paris*, Apr. 1, 1898; Léon Bourgeois, *Solidarité*, 1st ed. (Paris: A. Colin, 1896).

41. Joe S. Hayward, "The Official Social Philosophy of the French Third Republic: Léon Bourgeois and Solidarism," *International Review of Social History* 6 (1961): 20.

42. APSGM, dossier 317, letter from Edmond Bin to the president of the Syndicat du commerce, Aug. 29, 1901; letter from the Syndicat to Bin, Sept. 3, 1901; letter from Bin to Vve Pommery et fils, Aug. 3, 1902; letter to Bin from Pommery, Aug. 6, 1902; letter from Bin to Paul Krug, Aug. 18, 1902.

43. APSGM, dossier 317, letter to the Syndicat du commerce from Edmond Bin, Aug. 29, 1901; letter to Bin from the Syndicat, Sept. 3, 1901.

44. *Petit Parisien*, Oct. 15, 1908.

45. APSGM, dossier 317, letter to the Syndicat du commerce from Edmond Bin, Aug. 29, 1901; letter to Bin from the Syndicat, Sept. 3, 1901.

46., APSGM, dossier 317, letter from Edmond Bin to the president of the Syndicat du commerce, Aug. 29, 1901; letter from the Syndicat to Bin, Sept. 3, 1901; letter from Bin to Vve Pommery et fils, Aug. 3, 1902; letter to Bin from Pommery, Aug. 6, 1902; letter from Bin to Paul Krug, Aug. 18, 1902.

47. APSGM, Syndicat du commerce, Comptes rendus annuels, vol. 2, 1901–21, May 8, 1903.

48. For more on the proceedings of the Convention of Madrid, see AN, F11 2173, and *Journal Officiel*, Mar. 26, 1892.

49. APSGM, Syndicat du commerce, Comptes rendus annuels, vol. 2, 1901–21, May 8, 1903.

50. ADM, 169M27, letter from subprefect, Reims, to the prefect of the Marne, Nov. 11, 1904.

51. ADM, 155M22, "Réponses aux questions posées par le ministre de l'Agriculture, Syndicat agricole et viticole du canton de Vertus," 1901–2.

52. APSGM, dossier 317, letter from Edmond Bin to Vve Pommery et fils, Aug. 3, 1902; letter to Bin from Pommery, Aug. 6, 1902; letter from Bin to Paul Krug, Aug. 18, 1902.

53. APSGM, dossier 317, letter from Edmond Bin to Paul Krug, Aug. 18, 1902.

54. Quoted in Bourg-Broc, "Vocabulaire et thèmes politiques," 302.

55. ADM, 155M1, "Extrait du procès-verbal des délibérations du Conseil général," Apr. 20, 1903.

56. Ibid.

57. Ibid.

58. APSGM, dossier 317, letter from Edmond Bin to Paul Krug, Aug. 18, 1902.

59. ADM, 155M17, "Extrait du procès-verbal des délibérations du Conseil général," Aug. 20, 1903.

60. Ibid.

61. ADM, 155M1, letter to the prefect of the Marne from Paul Krug, Sept. 15, 1903.

62. ADM, 155M1, "Délibérations du Conseil municipal," Oct. 11, 1903; and "Récapitulation générale, répression de la fraude" (1903).

63. *Champagne viticole*, Dec. 1, 1903.

64. ADM, 155M1, minutes of meeting of the Comité pour la répression de la fraude, Nov. 10, 1903.

65. APSGM, Syndicat du commerce, Comptes rendus annuels, vol. 2, 1901–21, May 5, 1904.

66. Paul Vidal de la Blache, *Atlas général Vidal-la Blache* (Paris: A. Colin, 1909), 70.

67. André-Louis Sanguin, *Vidal de la Blache: 1845–1918 : Un Génie de la géographie* (Paris: Belin, 1993).

68. Quoted in Ibid., 121.

69. Guiomar, "Vidal de la Blache's *Geography*," 188.

70. Ibid., 189.

71. Ibid., 191

72. Quoted in Ibid., 204.

73. Paul Vidal de la Blache, *La France: Tableau géographique* (Paris: Hachette, 1908), 50.

74. "Des divisions fondamentales du sol français," *Bulletin littéraire* 2, no. 1 (1888): 1–7.

75. Sanguin, *Vidal de la Blache*, 295.

76. Paul Vidal de la Blache and P. Camena d'Almeida, *La France: Première A, B, C, D, Cours de géographie à l'usage de l'enseignement secondaire*, 15th ed. (Paris: A. Colin, 1918), 108.

77. Jacques and Mona Ozouf, "*Le Tour de la France par deux enfants*: The Little Red Book of the Republic," in Nora, *Realms of Memory*, 2: 126.
78. Ibid., 129.
79. Escoffier, *Souvenirs inedits*, 181.
80. See menus reproduced in Ibid., passim.
81. Spang, *Invention of the Restaurant*, 192.
82. See the discussion of the links between cookery and nation in ibid.
83. Étienne Clémentel, *Un Drame économique: Les Délimitations* (Paris: Pierre Lafitte, 1914), 89.
84. ADM, 155M1, letter from Conseil général to prefect of the Marne, Feb. 10, 1904; AN, F12 7001, memo of minister of agriculture on the Chambre de commerce de Saumur, Feb. 18, 1904. For the presentation of conflicting claims, see AN, F12 7001, "Rapport: Fraudes commerciales sur les vins," July 1904. For the debates in the Chambre des députés, see *Journal Officiel*, July 4, 10, and 12, 1904.
85. *Journal Officiel*, Feb. 24, 1905.
86. "La Crise viticole," *Temps*, Sept. 20, 1905.
87. APSGM, Syndicat du commerce, Comptes rendus annuels, vol. 2, 1901–21, May 3, 1906.
88. *Vigneron champenois*, Apr. 4, 1906.
89. Editorial by Miltat, vice president of the Fédération in *Vigneron champenois*, May 21, 1906.
90. *Éclaireur de l'Est*, Mar. 27, 1906; *Indépendant rémois*, Mar. 26, 1906.
91. *Éclaireur de l'Est*, Mar. 27, 1906.
92. Liszek, *Champagne*, 40–41, 46.
93. APSGM, dossier 117, pamphlet of the Syndicat commercial viticole champenois; ADM, Chp5665; Émile Moreau, *Histoire d'un syndicat: Les Mémoires d'un vigneron* (Aÿ: Syndicat commercial champenois, 1907).
94. ADM, E, *dépôt* 1504 F60, "Statuts, Syndicat des ouvriers champenois," 1906.
95. ADM, E, suppl. 1905, "Syndicat des ouvriers vignerons de Vertus." The organization was founded in 1905 and changed its name six years later.
96. ADM, Chp10203, Perrin, *Étude sur l'oeuvre des syndicats*.
97. APSGM, dossier 293, grape pricing. For reports on fraud, see, e.g., reports in *L'Illustration*, Jan. 12, 13, and 19, 1907.
98. ADM, 172M3, Chambre de commerce de Reims, commercial and industrial situation, 1906 (July 17, 1906).
99. ADM, E, *dépôt* 1504 F60, Association agéenne des propriétaires vignerons, 1906.
100. ADM, 155M2, petitions and letter from Fédération des syndicats de la Champagne to minister of agriculture, Aug. 5, 1906.
101. ADM, 155M2, letter from minister of agriculture to prefect of the Marne, Sept. 4, 1907.
102. *Figaro*, Sept. 5, 1908.
103. Armond Frémont, "The Land," in Nora, *Realms of Memory*, 2: 17.

104. Nora, *Realms of Memory*, 2: xi.
105. APSGM, dossier 25, Mercier.
106. ADM, 155M4, "Commission de délimitation de la Champagne viticole, séance du mardi 12 mai 1908."
107. "Protestation contre la convocation et la composition d'une commission de délimitation de la champagne viticole," in ADM, 155M4, "Commission de délimitation de la Champagne viticole, séance du mardi 12 mai 1908."
108. ADM, 155M4, "Commission de délimitation de la Champagne viticole, séance du mardi 12 mai 1908."
109. Ibid.
110. Ibid.
111. Ibid. Excerpts from the morning session, May 12.
112. Ibid.
113. Ibid.
114. Ibid.
115. Robert Gildea, *The Past in French History* (New Haven, Conn.: Yale University Press, 1994), 166–170.
116. ADM, 155M4, "Commission de délimitation de la Champagne viticole, séance du mardi 12 mai 1908."
117. APSGM, dossier 694, joint letter from the Syndicat du commerce and the Fédération des syndicats, Dec. 4, 1908.
118. *Matin*, Sept. 12, 1908. See also, ADM, 155M5, letter from prefect of the Marne to minister of the interior, Sept. 14, 1908.
119. ADM, 172M3, Chambre de commerce de Reims, document 11, commercial and industrial situation, 1908 (July 21, 1908; Feb. 9, 1909).
120. ADM, 155M5, "Extrait du procès-verbal des délibérations du Conseil général," Aug. 21, 1908.
121. *Matin*, Sept. 12, 1908.
122. ADM, 155M5, poster, "Appel aux *vignerons*" (1908) (emphasis in original).
123. ADM, 155M5, letter to prefect of the Marne, Sept. 24, 1908.
124. *Champenois*, Sept. 14, 1908; *Éclaireur de l'Est*, Sept. 15, 1908; *Indépendant rémois*, Sept. 14, 1908; ADM, 155M5, report from Commissariat spécial to prefect of the Marne, Sept. 24, 1908.
125. *Champenois*, Sept. 14, 1908; *Éclaireur de l'Est*, Sept. 15, 1908; *Indépendant rémois*, Sept. 14, 1908; ADM, 155M5, report from Commissariat spécial to prefect of the Marne, Sept. 24, 1908.
126. *Indépendant rémois*, Sept. 14, 1908.
127. ADM, Chp10203, Perrin, *Étude sur l'oeuvre des syndicats*.
128. APSGM, dossier 317, letter from Edmond Bin to Syndicat du commerce, Nov. 16, 1908; APSGM, dossier 317, reply from Syndicat du commerce to Bin, Nov. 25, 1908.
129. *Réveil*, Oct. 9, 1908; *Matin*, Oct. 27, Oct. 30, and Nov. 4, 1908.
130. *Éclaireur de l'Est*, Sept. 30, 1909. See also *Courrier de la Champagne*, Oct. 2, 1909.
131. *Matin*, Oct. 30, 1908.

132. *Petit Parisien*, Sept. 23, 1908; *Réveil de la Marne*, Oct. 9, 1908.

133. *Petit Parisien*, Oct. 15, 1908; *Réveil de la Marne*, Oct. 9, 1908.

134. ADM, Chp7065, Chambre de commerce de Reims, *Protestation* (cited ch. 2 n. 61 above).

135. Ibid.

136. *Champagne viticole*, June 1, 1910.

137. *Champagne viticole*, Aug. 9, 12, 23, 24, and 29, 1910.

138. APSGM, dossier 317, letter from Fédération to Syndicat du commerce, Feb. 13, 1909; reply from Syndicat du commerce to Fédération, Feb. 20, 1909. See also *Midi viticole*, May 18, 1909.

139. APSGM, dossier 694, "Affaire Péchadre."

140. AN, F12 6969, letter from Syndicat du commerce, June 1, 1909.

141. Jugement du Tribunal civil de Reims, Delouvin and Frappart, July 30, 1910; Jugement du Tribunal correctionel d'Épernay, Blondel affair, Nov. 7, 1910. See also reports of trials in *Réveil de la Marne*, Dec. 3, 27 1910; *Courrier de la Champagne*, Dec. 4, 1910

142. *Journal Officiel*, Feb. 27, 1910.

143. ADM, 172M3, Chambre de commerce de Reims, commercial and industrial situation, 1910 (July 12, 1910).

144. M. Lecacheur in *Champagne viticole*, Aug. 1910.

145. *Champagne viticole*, Sept. 1910.

146. *Réveil de la Marne*, Oct. 14, 1910.

147. Crowd estimates from Liszek, *Champagne*, 33.

148. AN, F7 13626, letter from prefect of the Marne to minister of the interior, Oct. 16, 1910.

149. ADM, 155M5, letter from subprefect, Épernay, to prefect of the Marne, Oct. 16, 1910.

150. Jean Nollevalle, "1911: Agitation dans la vignoble champenois," *Champagne viticole*, special issue, Jan. 1961: 10.

151. "La Guerre à la fraude," *Courrier de la Champagne*, Dec. 4, 1910; "Les Affaires des fraudes," *Réveil de la Marne*, Dec. 3, 1910.

152. APSGM, dossier 680, "Affaire Coste-Folcher"; *Réveil de la Marne*, Jan. 25, 1911, Jan. 29, 1911; *Courrier de la Champagne*, Jan. 29, 1911; *Éclaireur de l'Est*, June 28, 1911.

153. ADM, 155M5, report of subprefect, Épernay, to prefect of the Marne, Nov. 10, 1910. The seeming alliance between railroad employees and vignerons is discussed in Chapter 6.

154. ADM, 155M5, letters from subprefect, Épernay, to prefect of the Marne, Nov. 16, 17, and 18, 1910.

155. A. Nicolai, "La Délimitation de la region du bordelais," *Journal des Économistes* 1910, 80–92.

156. Paul Vidal de la Blache, "Regions françaises," *Revue de Paris*, Dec. 15, 1910, 812–49.

157. APSGM, dossier 694, joint letter of Fédération and Syndicat du commerce, Dec. 4, 1908.

158. APSGM, Syndicat du commerce, Comptes rendus annuels, vol. 2, 1901–21, May, 1906.
159. Nollevalle, "1911," 5.
160. In a promotional letter sent to me on Apr. 7, 1992, on the "origins and first activities of the syndicate," the Syndicat des grandes marques de Champagne gave all credit for legislative efforts and change in the region to the Syndicat du commerce and the négociants. Their "conscientious and diplomatic" actions, it stated, were the key to achieving protection of the regional appellation. Any mention of the need to protect against fraud within the industry is avoided. Letter in possession of author.

SIX 🍇 REVOLUTION AND STALEMATE

1. "La Champenois," sung to *L'Internationale*, Damery, Jan. 17, 1911 (reprinted in René Tys, "La Jacquerie champenoise de 1911," *Cahiers de l'Institut Maurice Thorez* 24 (1971): 124–27).
2. Loubère et al., eds., *Vine Remembers*, 120–21.
3. ADM, 155M6, report of subprefect, Épernay, to prefect of the Marne on his findings in Cumières, Jan. 17, 1911; note from the prefect of the Marne to minister of the interior on events in Cumières, Jan. 17, 1911; report from Captain Roland of Épernay on investigation of events of Cumières, Jan. 18, 1911.
4. ADM, 155M6, report of subprefect, Épernay, to prefect of the Marne on his findings in Cumières, Jan. 17, 1911.
5. ADM, 155M6. letter from prefect of the Marne to minister of the interior regarding the investigation of "manifestation," Jan. 17, 1911.
6. ADM, 155M6, letter from prefect of the Marne to minister of the interior regarding interrogation, Jan. 19, 1911; letter from prefect to minister of the interior regarding his trip to Venteuil, Jan. 19, 1911.
7. Nollevalle, "1911," 8.
8. Police reports provide no information on the motives of the protestors. ADM, 155M6 Captain Diez to subprefect, Reims, Jan. 18, 1911.
9. ADM, 155M6, confidential report of judge to subprefect, Reims, Jan. 20, 1911.
10. ADM, 155M6, report of Captain Diez to subprefect Reims, Jan. 28, 1911.
11. ADM, E, suppl. 2636, letter from subprefect to mayor of Cumières, Aug. 1, 1912. Damages at Cumières were estimated at 2,126,549 francs.
12. *Champenois*, Apr. 22, 1911.
13. "La Révolution en Champagne," *Réveil de la Marne*, Apr. 13, 1911.
14. "Lendemain de tragédie," *Humanité*, Apr. 14, 1911.
15. *Champagne viticole*, Sept. 1910.
16. *Courrier de la Champagne*, Apr. 20, 1889.
17. Patricia Prestwich, "Temperance in France," in *The Transformation of Modern France*, ed. William B. Cohen (Boston: Houghton Mifflin, 1997), 304.
18. ADM, 155M6, anonymous letter to subprefect, Reims, Jan. 18, 1911.
19. ADM, 155M6, letter from the négociant Michard Frères to prefect of the Marne, Jan. 20, 1911; report from subprefect, Reims, on conversation with the né-

gociant Jeannelle, Jan. 20, 1911; list of *points spéciaux* (n.d.); letter from subprefect, Reims, to prefect of the Marne regarding wine shipments at Gare des marchandises de Reims, Jan. 20, 1911.

20. Nollevalle, "1911," 9. Nollevalle did not doubt that the siege began on Jan. 20, 1911.

21. ADM, 155M6, letter from prefect of the Marne to minister of the interior, Jan. 24, 1911.

22. ADM, 155M6, report of commissaire spécial adjoint, Jan. 30, 1911.

23. APSGM, dossier 312, Montebello.

24. *Journal officiel*, Jan. 20, 1911.

25. Hervé Luxardo, *Les Paysans: Les Républiques villageoises, Xe–XIXe siècles* (Paris: Aubier, 1981), 27.

26. ADM, 155M6, report from captain of gendarmerie, Jan. 22, 1911, annotated in the margin by subprefect (emphasis in the original).

27. ADM, 155M6, report of subprefect, Reims, to prefect of the Marne regarding meeting in Aÿ, Jan. 20, 1911; Jan. 23, 1911.

28. ADM, 155M6, letter from prefect of the Marne to minister of the interior, Jan. 24, 1911.

29. ADM, 155M6, letter from subprefect, Épernay, to prefect of the Marne regarding meeting with négociants of Aÿ, Jan. 25, 1911.

30. ADM, 155M6, report from prefect of the Marne to minister of the interior regarding situation in the vineyards and requests of négociants to protect them, Jan. 24, 1911.

31. ADM, 155M6, letter on meeting with négociants from prefect of the Marne to minister of the interior, Jan. 25, 1911.

32. ADM, 155M7, Commissariat central de police, Feb. 5, 1911; see also *Union républicaine*, Jan. 28, 1911.

33. APSGM, membership bulletin of the Association syndicale des négociants en vins de la Champagne.

34. ADM, 155M7, Commissariat de police, report to subprefect, Reims, Feb. 3, 1911.

35. See ADM, 155M6, report of subprefect, Reims, to prefect of the Marne on *La Champagne commerciale*, Jan. 30, 1911.

36. See police reports in ADM, 155M7, Commissariat central de police, Feb. 3, 1911.

37. ADM, 155M7, open letter of négociants of the Syndicat de défense in *La Champagne commerciale*, Jan. 28, 1911.

38. "Protestation contre les mesures complémentaires," *La Champagne commerciale*, Jan. 28, 1911 (emphasis in original).

39. ADM, 155M7, open letter of négociants of the Syndicat de défense in *La Champagne commerciale*, Jan. 28, 1911.

40. "Notre but," *La Champagne commerciale*, Jan. 28, 1911.

41. *Union républicaine*, Jan. 28, 1911.

42. ADM, 155M6, prefect of the Marne to minister of the interior regarding accusations in the press made by négociants, Jan. 27, 1911.

43. ADM, 155M6, report of commissaire spécial adjoint, Jan. 30, 1911.
44. *Union républicaine*, Jan. 28, 1911.
45. ADM, 155M6, report of commissaire spécial adjoint, "La Situation à Venteuil / Détenteurs d'explosifs," Jan. 29, 1911.
46. J. Saillet, *Les Composantes du movement dans la Marne*, reprinted from *Mouvement social* 67 (Apr.–June 1969) (Paris: Éditions ouvrières, 1969), 81.
47. ADM, 152M7, letter on the "esprit des vignerons" from subprefect, Reims, to prefect of the Marne, Jan. 26, 1911; telegram from mayor of Verzenay to subprefect, Reims, Jan. 28, 1911; letter from subprefect, Reims, to prefect, Jan. 29, 1911.
48. *Journal Officiel*, Feb. 6, 1911.
49. *La Champagne commerciale*, Feb. 4, 1911; Mar. 12, 1911.
50. *La Champagne commerciale*, Feb. 17, 1911.
51. André Beury, *La Révolte des vignerons de l'Aube* (Troyes: Renaissance, 1984); Jacques Girault, "Le Rôle du socialisme dans la révolte des vignerons de l'Aube," *Mouvement social* 67 (Apr.–June 1969): 88–109.
52. Nollevalle, "1911 en Champagne," *Champagne viticole*, 11.
53. ADM, 155M8, "Quelques notes sur l'agitation dans la Marne" (n.d.); reprint of interview with Bertrand de Mun in *La Champagne commerciale*, May 6, 1911.
54. Loubère, *The Red and the White*, 318.
55. ADM, 155M7, letter from Fédération des syndicats de la Champagne to subprefect, Épernay, Mar. 15, 1911.
56. Loubère, *The Red and the White*, 318.
57. Girault, "Rôle du socialisme," 109.
58. Ibid.
59. *Journal officiel*, Apr. 5, 1911.
60. Ibid., Apr. 10, 1911.
61. "Au jour le jour," *Temps*, Apr. 3, 1911.
62. Nollevalle, "1911," 11.
63. ADM, 155M8, report on posters and copy of poster found by Captain Diez, Apr. 8, 1911.
64. Nollevalle, "1911 en Champagne," *Champagne viticole*, 11.
65. ADM, 2Z340 (B), letter from Commissariat de police, Apr. 1, 1911.
66. Ibid.
67. Ibid., "Livre noir des assassins de la Champagne" (Bibliothèque Sanglot, 1911).
68. ADM, 155M6, letter from subprefect, Épernay, to prefect of the Marne on meetings with cellar workers and Union coopérative des ouvriers cavistes et tonneliers d'Épernay, Jan. 26, 1911.
69. ADM, 197M7, reports of commissaire de police on partial strike at Union champenois, Mar. 28, 1911, meeting of cellar workers, Apr. 8, 1911, and strike of cellar workers in Épernay, Apr. 10, 1911; *Petit Temps*, Apr. 6, 1911; *Réveil de la Marne*, Apr. 12, 1911.
70. ADM, 155M6, list of *points spéciaux* (n.d.); ADM, 194M7, report of commissaire de police on strike of cellar workers in Épernay, Apr. 10, 1911.
71. ADM, 155M7, prefect of the Marne to minister of the interior, Mar. 30, 1911.

72. Isaie Richon, *Voix du Peuple*, Mar. 10, 1913.
73. ADM, 155M8, "Renseignements recueillis à Aÿ" (n.d.).
74. ADM, 155M8, confidential letter from minister of the interior to prefect of the Marne, Apr. 11, 1911.
75. ADM, 155M22, excerpt from telegram, 1911.
76. ADM, 155M6, "Journée du 12 Avril," 1911.
77. ADM, 155M6, report of General Abonneau, Apr. 21, 1911.
78. ADM, 155M22, complaint sent to *procureur* from Raymond de Castellance of Dizy-Épernay, Apr. 11, 1911 in.
79. ADM, 155M6, "Journée du 12 Avril," 1911.
80. ADM, 2Z15, report of Captain Tavernot to commanding general, Épernay, morning of Apr. 18, 1911.
81. ADM, 155M22, report of Captain Tavernot to commanding general, Épernay, afternoon of Apr. 18, 1911.
82. Nollevalle, "1911," 16.
83. Ibid., 15.
84. Ibid.
85. ADM, 155M8, "Démissions des municipalités du vignoble."
86. ADM, 2Z15, report of Captain Tavernot to commanding general, Épernay, afternoon of Apr. 18, 1911.
87. Nollevalle, "1911 en Champagne," *Champagne viticole*, 15.
88. ADM, 2Z15, report of Captain Tavernot to commanding general, Épernay, afternoon of Apr. 18, 1911.
89. ADM, 2Z15, telegram from Mareuil, 3:30 P.M., Apr. 12, 1911.
90. ADM, 155M8, report, Apr. 13, 1911.
91. *Réveil de la Marne*, Apr. 13, 1911.
92. ADM, 155M8, report, Apr. 13, 1911.
93. ADM, 155M8, report, Apr. 12, 1911.
94. ADM, 155M2, report of General Abonneau, Apr. 21, 1911.
95. *Réveil de la Marne*, Apr. 13, 1911.
96. *Temps*, Apr. 13, 1911.
97. ADM, 155M22, report of General Abonneau, Apr. 21, 1911, notes from telephone call to prefect of the Marne.
98. ADM, 155M8, mayor of Dizy to subprefect, Reims, Apr. 17, 1911; mayor of commune of Trépail to subprefect, Reims, Apr. 17, 1911.
99. Quoted in Luxardo, *Paysans*, 29.
100. ADM, 155M8, letter from Camuset to subprefect of Reims, Apr. 15, 1911; letter from Bouche fils to mayor of Mareuil, Apr. 28, 1911.
101. Ibid.; letter from Veuve Binet to subprefect, Reims, Apr. 15, 1911.
102. ADM, 155M8, letter from mayor of Verzenay to subprefect, Reims, Apr. 14, 1911.
103. *Humanité*, Apr. 26, 1911.
104. ADM, 155M8, letter from Captain Diez to subprefect, Reims, Apr. 27, 1911.
105. *Temps*, Apr. 14, 15, and 19, 1911.

106. *Journal des debats*, Apr. 24, 1911.
107. *Courrier du Champagne*, July 25, 1911.
108. Saillet, *Composantes du mouvement*.
109. ADM, 155M1, "Cour d'Assises du départment du Nord," Aug. 15, 1911.
110. Saillet, *Composantes du movement*, 81.
111. Ibid., 82–83.
112. Ibid., 81.
113. *Liberté*, Apr. 14, 1911; *Petit Temps*, Apr. 14, 1911; *Temps*, Apr. 14, 1911.
114. *Liberté*, Apr. 14, 1911.
115. ADM, 155M10, prefect of the Marne to minister of the interior regarding meeting of the Fédération, May 7, 1911.
116. ADM, 155M10, report of Captain Diez to subprefect, Reims, June 10, 1911; report of subprefect, Reims, to prefect of the Marne, June 28, 1911.
117. ADM, 155M10, commissaire spécial to subprefect, Reims, June 6, 1911.
118. Ibid.
119. AN, F12 6969, letter to minister of commerce from Comité en Champagne délimitée des vignerons ruinés, June 24, 1911.
120. ADM, 155M8, letter from Syndicat du commerce to Conseil général, Apr. 25, 1911.
121. *La Champagne commerciale*, May 6, 1911.
122. ADM, 155M1, quoted in report of commissaire de police, Aÿ, to subprefect, Reims, Oct. 25, 1912.
123. ADM, 155M12, letter from Commissariat spècial to prefect of the Marne, Feb. 24, 1912; letter from Commissariat spècial to subprefect regarding strikes in Aÿ, Feb. 26, 1912; letter from subprefect, Reims, to prefect of the Marne, Feb. 28, 1912.
124. Moreau, *Histoire d'un syndicat*.
125. AN, F7 13607, "Bourse du travail—Marne," and F713626, "Fédération et syndicats."
126. Saillet, *Composantes du movement*, 81–83.
127. Ibid., 81.
128. ADM, 155M14, letter from subprefect, Reims to prefect of the Marne, Feb. 7, 1913.
129. ADM, 155M13, report to minister of labor on the strike at Hautvillers, Feb. 1912.
130. ADM, 155M13, report on vigneron strike in Verzenay, Mar. 1912.
131. ADM, 155M13, report from subprefect, Reims, to prefect of the Marne on meeting in Aÿ, Oct. 25, 1912.
132. ADM, 155M13, letter from subprefect, Reims, to prefect of the Marne, Mar. 7, 1912.
133. ADM, 155M13, letter from subprefect, Épernay, to prefect of the Marne, Aug. 28, 1912.
134. ADM, 155M14, letter from prefect of the Marne to subprefect, Reims, Feb. 6, 1913. Similar concerns were voiced about the region around Reims. ADM, 155M14, letter from subprefect, Reims, to prefect of the Marne, Feb. 14, 1913.

135. ADM, 155M14, letter of subprefect, Reims, to prefect of the Marne, Feb. 14, 1913. "The *grandes Maisons*," he wrote, "to use the expression adopted in the vineyards, maintain that the petits propriétaires vignerons are behind the strikes."

136. APSGM, dossier 680, "Statuts, 1913."

137. ADM, 155M14, letter from subprefect, Reims, to prefect of the Marne, Feb. 7, 1913; *Progrès de l'Est*, Nov. 12, 1913.

138. For a discussion of the links between these two strands of nationalism, see Zeev Sternhell, "The Political Culture of Nationalism," in *Nationhood and Nationalism in France*, ed. Robert Tombs (New York: Harper Collins, 1991), 22–38.

139. Ibid.

140. ADM, 155M12, letter from subprefect, Reims, to prefect of the Marne, Feb. 20, 1912.

141. Malherbe, "Socialisme et vignoble dans la Marne"; Liszek, *Champagne*.

142. Isaie Richon, *Indépendant rémois*, Aug. 10, 1911.

143. ADM, 155M1, quoted in the report of commissaire de police, Aÿ, to subprefect, Reims, Oct. 25, 1912.

144. *Observateur de l'Est*, Feb. 6, 1904.

145. "Appel aux vignerons," *Independant rémois*, Feb. 23, 1905.

146. Moreau, *Histoire d'un syndicat*.

147. "La Révolution en Champagne," *Réveil de la Marne*, Apr. 13, 1911.

148. "La Guerre à la fraude," *Courrier de la Champagne*, Dec. 4, 1910.

149. James Lehning, *Peasant and French: Cultural Contact in Rural France during the Nineteenth Century* (New York: Cambridge University Press, 1995), 11.

150. Weber, *Peasants into Frenchmen*, 486.

SEVEN 🍇 CONCLUSION

1. Hodez, *Protection des vins de Champagne*, 320.

2. Semaine nationale du vin, *Compte-rendu* (cited ch. 2 n. 132 above), 378.

3. For an excellent study of the importance of the new transport culture in the twentieth-century and cuisine, see Stephen L. Harp's *Marketing Michelin: Advertising and Cultural Identity in Twentieth-Century France* (Baltimore: Johns Hopkins University Press, 2001).

4. Crestin-Billet, *Naissance d'une grande maison du Champagne*, 187.

5. See Pierre Bourdieu, *Distinction: A Social Critique of the Judgment of Taste* (Cambridge, Mass.: Harvard University Press, 1984), 53.

6. Escoffier, *Souvenirs inedits*, 191.

7. For an excellent discussion of the place of "the land" in French memory, see Frémont in Nora, *Realms of Memory*, 2: 17.

8. Nora, *Realms of Memory*, 2: xi.

9. Peter Sahlins, *Boundaries: The Making of France and Spain in the Pyrenees* (Berkeley: University of California Press, 1989), 7.

10. See, e.g., Weber, *Peasants into Frenchmen*; Karl Deutsch, *Nationalism and Social Communication: An Inquiry into the Foundations of Nationality* (Cambridge, Mass.: Harvard University Press, 1953).

11. For critical reviews of the reality and timing of this shift, see Charles Tilly, "Did the Cake of Custom Break?" in *Consciousness and Class Experience in Nineteenth-Century Europe*, ed. Merriman, 17–41, and Ted W. Margadant, "French Rural Society in the Nineteenth Century: A Review Essay," *Agricultural History* 53 (1979): 644–51.

12. Sahlins, *Boundaries*, 9.

13. For more of an overview of the literature on the oppositional model of national identity, see Sahlins, *Boundaries*, 9–10.

14. Benedict Anderson, *Imagined Communities: Reflections on the Origin and Spread of Nationalism* (London: Verso, 1983), 15. This definition of nation as community is one way to focus attention on the symbolic construction of national identities.

15. Olson, "Tempest (but No Bubbles)."

Bibliographic Essay

Wine inspires. Historians, much like artists, philosophers, and writers, have used this inspiration to capture the merits and magic of the ancient elixir. Accordingly, there are shelves of books overflowing with well-aged anecdotes about champagne's paternity flowing from the blind monk Dom Pérignon discovering bubbles in still wine at the abbey of Hautvillers in the old province of Champagne. To add complexity, historians discuss the seemingly timeless production methods used for making bubbly. Champagne is rarely associated with modernity or industrial capitalism; it is always, however, French. The wine, much like the rural community that produced it, is ensconced in the imagery of tradition and a nostalgia for an idyllic French countryside. Historical narratives are inevitably grounded in a discussion of the components of terroir, enumerated as scientific fact independent of social construction. Few scholars question *when* or *how* sparkling wine came to be symbolically identified as the quintessential wine of France connected with celebration and fraternity. None investigate what this investiture with cultural capital meant for those in rural France who produced it.

There are, nonetheless, a variety of books on champagne and the wine industry more generally that can provide a useful starting point for those interested in the history of French wine. Roger Dion's *Histoire de la vigne et du vin en France des origins au XIXe siècle*, first published in 1959, discussed at length in the Introduction, remains the classic work on the French wine industry. Dion examines the interplay between geography and human activity and its influence on wine production from the Gallo-Roman era until the French Revolution and the "glorious ascension of Champagne." The sweep is breathtaking. Wine, the "subject of France's collective pride," in Dion's words, is treated with great reverence as part of the *patrie française*, and a comprehensive bibliography attests to the extent to which other French writers share Dion's fascination with it.

English-language writers and historians outside of France have equally contributed to our collective knowledge of champagne and the wine industry. Books by wine experts and journalists tend to dominate the literature. There are surprisingly few scholarly works on the wine industry, and no recent works on the history of the champagne wine industry in the nineteenth and twentieth centuries. Rod Phillips's *A Short History of Wine* (2000), Tim Unwin's *Wine and the Vine: An Historical Geography of Viticulture and the Wine Trade* (1991), and Leo Loubère's *The Red and the White: A History of Wine in France and Italy in the Nineteenth Century* (1985) and *The Wine Revolution in France* (1990) are among the best-known general histories of wine in English. All four works seek a sweeping overview of the French wine industry, but offer limited coverage of the sparkling wine industry. Phillips's work is the most recent and the most accessible to the general reader.

While not focused exclusively on France, Phillips goes further than any other historian in consciously recognizing wine as a marker of French national identity and describing the link between the two. Unwin provides a broad overview of the international wine trade from antiquity to the present from the perspective of historical geography. His work is particularly focused on the interplay between culture and physical environment, a standard concern in the discipline of geography. Using the methods of history rather than historical geography, Loubère has produced two very different books. He provides excellent coverage of the French still wine industry in both the nineteenth and twentieth centuries, but his discussion of the sparkling wine industry in both works is, like those of Unwin and Philips, based mainly on published secondary sources and relatively limited.

In *The Red and the White,* Loubère notes that most works in French and English on champagne "subsume fact and fancy, legend, and propaganda." Little has changed in the two decades since he wrote these words. Indeed, Loubère's observation seems particularly appropriate for the few books published on champagne since the 1980s. There have been popular works, such as Frédérique Crestin-Billet's *La Naissance d'une grande maison de Champagne: Eugène Mercier, ou l'audace d'un titan* (1996), written by journalists hired by the industry or underwritten and published by individual wine houses. Few of these works discuss the vine growers, focusing attention instead on wine or individual wine houses. Many of these works on wine houses are written by descendents of their founders, such as Prince Alain de Polignac's *Madame Pommery: Le Génie et le coeur* (1994). These works have no scholarly pretenses and generally rely, as Loubère once noted, on the "memories of old men." François Bonal, a retired French colonel turned local historian, has published a number of books, including *Le Livre d'or du champagne* (1984) and *Dom Pérignon: Vérité et légende* (1995), that have attempted to chronicle and occasionally dismantle some of the "legends and propaganda" in the prevailing literature.

One of the most notable books on champagne in recent years is Thomas Brennan's masterful *Burgundy to Champagne: The Wine Trade in Early Modern France* (1997). Brennan examines the development of the French wine trade in the early modern period, arguing persuasively that it was central to the commercialization of France's rural economy. Moving away from the heavy emphasis on the relationship of geography, environment, and human efforts in elaborating wine, Brennan examines the growth of markets and marketing networks. His study of champagne ends around 1810, before champagne began its transition to an "industry," complete with factories and assembly-line production. My work picks up the narrative where Brennan leaves off by tracing the transition to a mass consumer-oriented industry serving the new culture of food so brilliantly described and analyzed by historians such as Rebecca L. Spang (*The Invention of the Restaurant* [2000]) and W. Scott Haine (*The World of the Paris Café* [1996]).

The extraordinary relationship between the French and cuisine, as these recent works attest, has not gone unnoticed by scholars. Serious scholarly investigation of diet, ingredients, and rituals of consumption has progressed rapidly since its beginnings with practitioners of the *Annales* school in the 1970s. Purely quantitative methods, favored by some of these early practitioners, gave way to looking at cultural contexts by the 1980s. Historians influenced by the work of cultural anthropologists and ethnographers, such as Sidney W. Mintz in *Sweetness and Power: The Place of Sugar in Modern History* (1985), began to explore the social importance of food and rituals of food and wine consumption. Sites of consumption, such as restaurants, cafes, and public banquets have yielded new insights into how food and wine serve in the construction of social identity. Despite these wonderful contributions to our understanding of food and wine, historians still understand very little about how wine was invoked at the local and national levels as a part of "Frenchness." Eating and drinking, with their daily repetition, may appear insignificant when placed next to the great deeds and events of history. Yet, historically, wine has been a central preoccupation in France, with a significance that goes well beyond simple acts of daily consumption.

The history of champagne is not exclusively a story of wine and culinaria. This book has been informed by current debates about the links between material culture and identity. In the past decade the most influential way of explaining the sense of national belonging has been to regard the nation as a cultural artifact, a product of "invented traditions" (Eric Hobsbawm and Terence Ranger, *The Invention of Tradition* [1983, 1992]) and "imagined communities" (Benedict Anderson, *Imagined Communities: Reflections on the Origin and Spread of Nationalism* [1983, 1991]). There has been a flourishing interest among scholars and policymakers in nationalism, identity, and the creation of national attachments. Sharp disagreements have emerged, not only over the nature of French national identity, but also over the extent to which diverse and sometimes hostile provincial communities were integrated into the nation. As James Lehning notes in *Peasant and French* (1995), "Imagining the French community has meant determining both who would be included in the nation and what form that inclusion would take" (p. 11). It has also meant a variety of ways of imagining the place of the rural population in the French nation. Generations of historians saw rural society as having eternal qualities linked to the land and its people. The task of the French nation in the late nineteenth century, when wines like champagne were reaching new levels of consumption, was to integrate the separate *pays* that made up the rural world into the nation in a process that Eugen Weber calls "akin to colonization" in his book *Peasants into Frenchmen* (1976).

While the material goods of rural France were being embraced for their embodiment of ancient French tradition and civilization, those who produced them in the provinces, it is suggested, were the focus of a drive to civilize on the part of the Third Republic. In order to construct the nation, a backward countryside had to be assimilated into the more homogeneous, urban-based world. Scholars generally agree that center-driven phenomena of urbanization, mass education, secularization, and modern forms of market organization linked to the development of industrial capitalism helped to generate the idea of "France" as nation. But the extant to which this construction of the nation was then imposed by Paris on the periphery is open to debate. Recent work, from Caroline Ford's study of Brittany in *Creating the Nation in Provincial France* (1991) to Peter Sahlins's pathbreaking examination of the French frontier in *Boundaries: The Making of France and Spain in the Pyrenees* (1989), have demonstrated the need for a reexamination of the center-based typology of nation formation, a reexamination of the relationship between center and periphery. These works argue that nation formation involves much more negotiation, accommodation, and resistance than previously suggested. Moreover, a new emphasis on the importance of material goods to the creation of collective identity and French national culture in the works of historians such as Leora Auslander (*Taste and Power: Furnishing Modern France* [1996]), Whitney Walton (*France at the Crystal Palace: Bourgeois Taste and Artisan Manufacture in the Nineteenth Century* [1992]), and Debora Silverman (*Art Nouveau in Fin-de-Siècle France: Politics, Psychology, and Style* [1989]) suggests that the importance of agricultural goods, such as wines, and the terroir in which they are produced needs to be reconsidered. Champagne and other areas of rural France became the sources of some of the nation's most recognized and prized material goods in the nineteenth and twentieth centuries. Indeed, as discussions of food and wine in Pierre Nora's influential compilation *Les Lieux de mémoire* (1984–86), translated by Arthur Goldhammer under the title *Realms of Memory: The Construction of the French Past* (1996–88), particularly volume 2, *Traditions,* suggest, wine, food, and land became a pivotal part of French collective memory in the modern period.

The sources for examining champagne as a part of collective memory and material culture were not difficult to come by. Champagne is ubiquitous in both popular and elite representations. Reflecting its place in the emerging consumer culture of the late nineteenth century, champagne appeared with frequency in art and literature, and it is the focus of paintings such as Édouard Manet's *Un Bar aux Folies-Bergère* and Paul Cézanne's *Chaise, bouteille et pommes*. Poster art from the turn of the century featuring champagne was created by artists as diverse as Pierre Bonnard, Alphonse Mucha, Walter Crane, Leonetta Cappiello, and Henri Toulouse-Lautrec, and is still widely recognized for its artistic merits. Champagne took on symbolic importance in the writings of Émile Zola (*Nana*), Alexander Pushkin (*Eugene Onegin*), and Johann von Goethe (*Faust*). Indulgence in the bubbly elixir also found a place in the literature of the popular classes. In *La Vie parisienne*, *Rire,* and *Punch,* humorists like Honoré Daumier, John Leech, and Richard Voigts were inspired to use champagne as part of their satire of both elite and middle-class culture. And in music from the era, champagne was integral to popular beer-hall and café tunes, such as "Champagne Charlie" and the "Ruinart-Polka," as well as orchestral pieces such as the "Charles Heidsieck Waltz" by Paul Mestrozzi, which debuted at a ball for the Austrian emperor on January 26, 1895. Champagne was consumed as both wine and image.

The champagne manufacturers were often interested in representations of their products. Letters and memos from négociants, the Syndicat du commerce, and the Reims Chamber of Commerce to the minister of agriculture and the minister of commerce housed in series F10 and F12 in the Archives nationales in Paris attest to how closely these images were surveyed by négociants and their representatives. Traveling sales staff working for champagne firms, as well as representatives of the national ministries themselves, provided frequent updates on popular representations. Particular attention was focused, however, on commercial usage of the name "champagne" and the counterfeiting and imitating of established French brands. The close cooperation between private industry and the French government evident in these dossiers is striking.

Private archives are indispensable for the study of the attempts both to control these popular representations of champagne and to create images that would attract consumers. Many private company archives still remain closed to researchers, and access to documents in those that are open is sometimes limited. A few firms, such as that of Krug, have deposited their business records in the series J of the Archives départementales de la Marne in Châlons-en-Champagne. Researchers who want to go beyond this limited selection are required to address themselves to each firm. For this study, I found the private archives of the Syndicat des grandes marques de Champagne in Reims, the current incarnation of the nineteenth-century Syndicat du commerce, to be particularly rich. Hundreds of dossiers on local, national, and international issues concerning champagne have been carefully preserved from the earliest days of the Syndicat. Members of the organization often sent the director letters with newspaper articles, editorials, and ads attached for comment. Their focus was both international and national, and one finds dossiers on the Miller Brewing Company of Milwaukee—the maker of the "champagne" of beers—alongside dossiers on sales to Russia. Information on counterfeiting cases abroad are fully documented, as are the legal cases against fraudulent producers in France. And, since this was the promotional organization for the product category of sparkling French wine as well as local producers, there are dossiers containing Syndicat-created promotional materials. Moreover, the organization kept invaluable records on local wine production, correspondence with local peasant organizations, minutes from organizational meetings, correspondence with national leaders, and documents concerning international accords.

The négociants had a direct need to communicate effectively with their audiences of relatively select customers in Europe and the United States who could afford this expensive, luxury commodity. Without modern ad agencies and the use of consumer surveys and questionnaires, négociants experimented with advertising appeals through packaging of the commodity (in this case through wine labels). What is striking about these champagne wine labels is that, unlike modern labeling, they tended to include elaborate pictures and written texts, signs meant both to "instruct" the consumer and to appeal to the individual client. A public register of these wine labels was created in 1859 to record the trademarks and labels used by the Champagne négociants and their firms. Firms were not required to register their labels or trademarks, but many took advantage of the recording mechanism, which gave them a legal advantage if they had to pursue a counterfeiting case in court. Hundreds of labels, now carefully preserved in the Archives départementales de la Marne series 16U, were entered into the register between 1859 and 1902. Although not exhaustive, this record, combined with surviving promotional pamphlets and posters scattered in the Bibliothèque nationale de France in Paris, the Bibliothèque municipale in Châlons-en-Champagne, and the

Carnegie-endowed Bibliothèque municipale in Reims provides a substantial data base for analyzing marketing trends.

Understanding symbolic expressions within champagne advertising and promotional materials was made easier by employing methods pioneered by historians since the 1980s. For help in deciphering the political meaning, particularly of early nineteenth-century prints, I found two works particularly helpful: Lynn Hunt's *Politics, Culture, and Class in the French Revolution* (1984) and Serge Bianchi's *La Révolution culturelle de l'an II: Élites et peuple, 1789–1799* (1982). One useful survey of the importance of revolutionary iconography and republican imagery after the French Revolution was Maurice Agulhon's *Marianne into Battle: Republican Imagery and Symbolism in France, 1789–1880* (1981). Central to my overall analysis of champagne marketing, however, are the works of Thomas Richards, *The Commodity Culture of Victorian England*; Marc Martin, *Trois siècles de publicité en France* (1992); and Aaron Jeffrey Segal, "The Republic of Goods: Advertising and National Identity, 1875–1918" (Ph.D. diss., University of California at Los Angeles, 1995).

Analyzing the group of négociants who produced these documents remains more elusive. Historians have left journalists (sometimes hired by the champagne firm) and descendents of wine houses to study the business of champagne in the nineteenth century. While some of these studies cite original documents from private archives, the authors tend toward hagiography, as titles such as Crestin-Billet's *Eugène Mercier, ou l'audace d'un titan* (Eugène Mercier, or, The Audacity of a Titan) and de Polignac's *Madame Pommery: Le Génie et le coeur* (Madame Pommery: The Genius and the Heart) suggest. Even less is known about business structures, business politics, and institutional procedures of the champagne firms. Thomas Brennan's study of the early modern wine trade is one of the rare scholarly monographs that makes use of extensive private business archives. In an image-conscious industry, firms are much more reluctant to share information on the modern era, particularly around the period of the 1911 riots. As Leo Loubère has noted, the archives "have either been destroyed during war, or remain closed, or, where opened, contain little information" (*The Red and the White*, 246).

In recent years, scholars have either implicitly or explicitly questioned the utility of studying individual business enterprises or business leaders to address broader questions of social and economic change. My study follows these trends, moving away from an emphasis on the individual entrepreneur or enterprise. I focus on commodity production within a community and thus look at the production of raw materials as well as the industrial transformation of those raw materials. In short, I examine the overlapping worlds of négociants and the vignerons, the worlds of both agriculture and industrial food production. Much of the existing literature on the champagne industry suggests that the formative impulse for change within the champagne industry, particularly regarding appellations and delimitations, came from the négociants, with the more or less docile support of the vignerons. Petitions and letters from the prefect of the department of the Marne found in the dossiers of the minister of agriculture in the Archives nationales in

Paris suggested that the vignerons were vocal about issues affecting the industry. It was in the Archives départementales de la Marne that I discovered just how active they were in industry affairs.

Peasant activism can most clearly be seen in regional newspapers. The Archives départementales de la Marne has full runs of all regional newspapers. Not only are the activities of peasant organizations covered in the papers such as the *Courier de la Champagne*, *Éclaireur de l'Est*, *Réveil de la Marne*, and *Indépendant rémois*, but there are specific papers dedicated to vineyard affairs. *La Champagne commerciale*, *La Champagne viticole*, and *Le vigneron champenois* proved to be rich sources of information on local events and opinions. The prefect also kept files of newspaper clippings from national and international papers concerning peasant activity and industry concerns. These are carefully filed in the M series of the same archives. Finally, the short-lived paper of the activist René Lamarre, *La Révolution champenois*, was an invaluable source of information on syndicate organization and efforts at creating cooperatives.

Because grape growing and wine production were so central to the economic, social, and political life of the region, the Archives départementales de la Marne has cartons bulging with thousands of pages of material on the subject. I found the bureaucratic records of the prefect in the series M to be enormously helpful. Letters, memos, and reports regularly flowed to and from the prefect's office and the various ministries in Paris, the mayors of local communes, police chiefs, and private citizens. At critical points during this period, assessments of landholding patterns in each commune were compiled and sent to the prefect. Moreover, this series contains hundreds of dossiers on matters both large and small—from runaway donkeys to syndicat membership—of concern to the rural community. Series E and Z, containing records of individual communes, offer another layer of documentation on the rural community, although sometimes overlapping with series M. There were occasional rare finds to be made here, such as the hand-written "Livre noir des assassins de la Champagne" (Black Book of the Assassins of Champagne), picked up by the police in the vineyards, in series Z. Finally, of course, police records (mainly mixed in series M) of the riots and surveillance of meetings and the judicial records in series U are indispensable for understanding the uprisings surrounding the phylloxera and, again, in 1911.

This study has noted the importance of gender in both the creation of the image of champagne and the development of the industry. Veuve Clicquot, Veuve Binet et fils, Veuve Loche, Veuve Pommery—these names among the list of prominent wine houses suggest that women played an important role in the development of the champagne industry. Widows and married women, whose names appear in the membership rolls of peasant syndicats, were active participants in the history of the industry and region. Their stories, however, are not always easy to uncover in the public record. "A genuine mother, attentive and devoted" became the classic description applied to the "grandes dames du Champagne" who were so prominent in the champagne industry during the nineteenth century. Similar descriptors were used for women who labored in the vineyards. While the voices of individual women are often hard to discern in public or private records, their collective involvement in the business of winemaking is abundantly clear. My study of the champagne industry suggests that French wine was not, as it is so often portrayed, a bastion of male prerogative. Looking at the history of the champagne industry reminds us of the gendered nature of the historical record. Historians have only begun to examine how women, like wine, were incorporated into notions of "Frenchness."

Index

Abbey of Hautvillers. *See* Hautvillers, Abbey
Abonneau, General, 174–75
Académie des sciences, 43, 90, 91
Académie française, 42
advertising. *See* marketing of champagne
"Affaire Coste-Folcher, L'," 154
Agathe, Saint, festival of, 66
Aisne, department of, 93, 118, 119, 147, 149, 161, 168, 192
Albert, Marcelin, 152, 175
alcoholism, 31–32
Alsace, 63, 83, 187
Ambonnay, 55, 197
American rootstock, 87, 90–91, 95, 107, 108, 114, 115, 187
ancien régime, 10, 14, 149, 191
Anjou, 121
AOC laws. *See* appellation d'origine contrôlée laws
aphids. *See* phylloxera
appellation d'origine contrôlée (AOC) laws, 8, 121, 146, 186, 189, 192–93, 195
Ardennes, department of, 5, 133
Armagnac, 150, 155
Aron, Jean-Paul, 19, 30
Association des vignerons champenois, société coopérative de production, 116
Association syndicale autorisée pour la défense des vignes contre le phylloxéra: controversy surrounding, 102–4, 106–7, 110–12; formation of, 87–88; failure of, 114–15
Association syndicale des négociants, 181
Association syndicale des propriétaires-vignerons de Damery, 105
Association viticole champenois (AVC), 114–15, 117, 127, 152
Aube, department of, 5, 48, 52; and delimitation battles, 118, 119, 133, 146–49, 161, 166–69, 177, 184, 192
AVC. *See* Association viticole champenois
Avenay, 197
Avize, 25, 54, 58, 82, 111, 197
Aÿ, 28, 48, 80, 97, 103, 142, 166, 197; prestige of, as a reputable grape-producing area, 27, 55, 144; price of grapes in, 82; and revolt of 1911, 164, 171, 173–77; survey of landholding in, 56; vigneron meeting in, 150
Ayala firm, 174

Balzac, Honoré de, 21, 43
Banyuls, 150, 186
Barrès, Maurice, 173
Bar-sur-Aube, 169
Basse Montagne, 197
Bazille, Gaston, 89
Beaujolais, 48
Belle Époque, 31–33, 44, 119, 128, 145
Bergson, Henri, 136
Berthet firm, 159, 162
biens nationaux, 47, 48
Bin, Edmond, 128–32, 142, 143, 151, 157
Binson-Orquigny, 158
Bissinger firm, 174
"Black Book of the Assassins of Champagne." *See* "Livre noir des assassins de la Champagne"
Blaye, 145
Bloc des gauches, 131
Blum, Léon, 136
Bollinger firm, 40, 84
Bolo-Pacha, 152–54
Bonal, François, 59, 63
Bonaparte, Napoleon, 14–15
Bonnard, Pierre, 33
Bonnedame, Raphaël, 118–19, 123, 136
Bordeaux (region), 50–51, 152; classification system for, 136; and delimitations, 150, 155, 161, 166; importance of terroir in, 80; study of military garrison in, 43; traditions of, 72; wine merchants in, 20
bordeaux (wine), 186, 192, 193
Bordeaux Conference of 1881, 91, 95
Boulanger, Georges, 67
Bourdieu, Pierre, 5, 188
Bourgeois, Eugène, 165
Bourgeois, Léon, 67, 116, 129, 130, 135, 144, 169
Boursault, 17, 109, 158

239

bourse de travail (labor exchange), 57, 183
Bouzy, 27, 54, 55, 61, 64, 97, 197, 210n. 144; prices for grapes in, 58, 82; red wine of, 49
Brennan, Thomas, 4, 14
brie, 19, 44, 192
Brillat-Savarin, Anthelme, 1, 42–43
Brisson, Adolphe, 1
Brittany, 67
Broyes, 112
Bruno, G. *See* Fouillée, Augustine
Burgundy, 10, 51, 56, 62, 68–69, 121–22
Burgundy to Champagne: The Wine Trade in Early Modern France (Brennan), 4

cahiers de doléances, 47, 48
Cappiello, Leonetto, 33
carbon bisulfide treatment, 91
Castellane firm, 172
Catholic Church, 29, 47, 65
Catholic monasteries, 47–48
Cazanove, Charles de, 25
CGT. *See* Confédération générale du travail
Châlons-en-Champagne, 51
Châlons-sur-Marne, 51, 52, 133, 135, 147
Chamber of Commerce, 106, 107, 128, 143, 151, 153, 234; conflicts within, 76–77; documents of, 167
Chamber of Deputies, 132, 144, 154, 166, 167, 169
Champagne commerciale, La, 165, 167–68, 177, 237
Champagne viticole, La, 134–35, 143, 153, 237
Chandon de Briailles, Gaston, 74, 94, 95, 105
Chandon de Briailles, Paul, 92–93
Chandon de Briailles, Raoul, 56, 65, 103, 142, 175
Charente, 107, 132
Charles X, 14
Châteaux Margaux, 27
Châteaux-Thierry, 149, 197
cheese, French, 189, 192. *See also* brie; roquefort
Chevigné, comte de, 17
Chigny-les-Roses, 197
cholera, 53
Chouilly, 55
CIVC. *See* Comité interprofessional du vin de Champagne
Clairvaux, 147

Clicquot family, 14, 15, 17, 23, 73. *See also* Veuve Clicquot–Ponsardin (firm)
cochylis moths, infestation of, 81
cognac, 19, 44, 132, 150, 155, 192
Comice agricole et viticole d'Épernay, 92–93
Comité des vignerons ruinés de la Marne, 170, 175, 177
Comité d'étude et de vigilance contre le phylloxéra. *See* Vigilance Committee
Comité interprofessional du vin de Champagne (CIVC), 8
Commission supérieur du phylloxéra, 90
Condé-en-Brie, 149
Confédération générale agricole pour la défense des produits purs, 152, 153
Confédération générale du travail (CGT), 171, 179, 183
Conseil d'État, 141, 150, 154
Conseil général, 95, 106, 131, 133, 135, 141–42, 153, 177
Convention of Madrid, 120–21, 131
Convention of Paris of 1883, 96
corking, 70
Côte d'Avize, 111, 197
Côte d'Épernay, 197
Côte des Blancs, 68, 73, 197
Coutant, Paul, 116
Cramant, 54, 111, 197
Cumières, 109, 113, 158, 197

Damery, 40, 54–55, 112; and peasant resistance, 155, 158, 163, 172, 176, 178; producers' cooperative established in, 111, 116; and René Lamarre, 40, 99, 108, 109, 143
Dargent family, 48
Daumard, Adeline, 17, 33
Delanoue, Thierry, 147, 148
Delouvin firm, 165, 172
departmental council. *See* Conseil général
depression: of the 1880s, 80–85, 125; of the 1930s, 192
Déroulède, Paul, 173, 187
Dervin, Abbé, 94
Deutz & Geldermann, 25, 73, 104, 173, 174
Die, 155
Dion, Roger, 2–3, 15, 231
Dizy, 155, 172, 197
Dormans, 101
Doutté, Edmond, 95, 99, 107, 110
Dreyfus Affair, 34, 67, 84, 104, 131

École normale supérieure, 136
Egypt, 19, 152
Elias, Norbert, 17
England. *See* Great Britain
enology, science of, 136
Épernay, 19, 54–56, 61, 66, 97, 103, 110, 116, 123, 143, 165, 197; antipathy to newcomers in, 20; damage to vines in, 153; Moët & Chandon in, 70; peasant resistance in, 133, 135, 142, 154; and revolt of 1911, 159, 164, 166, 168, 169, 171–76; socialists in, 84; statue of Dom Pérignon in, 28; success of wine merchants in, 13, 15
Escoffier, Auguste, 91, 139, 140, 189
European Union, 2, 121, 193
Exposition universelle: in Brussels (1910), 29; in Paris (1878), 28, 35; in Paris (1889), 37

Fédération des syndicats viticoles de la Champagne, 129, 132, 161; attempts at delimitation by, 135, 141–42, 149–54, 156–57, 163, 169–70, 172, 175–77, 180–83; dealings of, with the Syndicat du commerce, 130–31; founding of, 127–28; leadership of, 143
Figaro, Le, 144, 145, 146
Fleury, 109, 158
Fontaine, Paul, 165
Ford, Caroline, 67
Fouillée, Augustine, 139
Fournier, Jules Félix, 18
fournisseurs brevetés (purveyors by appointment), 18
France, La (Vidal de la Blache), 139
Franco-Prussian War, 63, 77, 84
Franco-Russian alliance of 1893, 34
François I, 40
fraud in the wine industry, 7, 98–99, 118–57, 162, 166, 169–70, 172, 177, 184
French Revolution, 7, 10–11, 14, 28, 155, 181, 231; anniversary of, 34, 82; legacy of, 5, 47–48, 118, 148–49, 161, 184, 190–91; period before, 13, 42; as a symbol, 79, 173, 236
fusée paragrêle (hail rocket), 60, 166

G. H. Mumm firm. *See* Mumm, G. H.
Gaillard, Maxime, 31
Gallois firm, 174
Gardet firm, 171

gastronomy, 1, 19, 20, 30–31, 42–44, 139–40, 232–33
Gauthier firm, 171, 174
Geertz, Clifford, 19
geography, 43–44, 136–39, 155, 188, 191, 192, 231, 232
Germans/Germany, 23, 33, 37, 79, 99, 138; during World War I, 152; in French champagne industry, 22, 73; as producers of fraudulent champagne, 80, 99, 121, 146, 173; rhetoric condemning, 83–85, 104, 117, 122, 130, 170, 181
Gide, Charles, 84, 125
Giesler firm, 22, 73, 75, 77
Gillis, John, 34, 35
Gironde River, 146
Goffman, Erving, 32
grafting of vines, 62, 90–91, 95, 114. *See also* American rootstock
grande marque, definition of, 24
Grande Pharmacie, 32
Great Britain, 14, 22, 36, 37, 97, 131
Griffe law of 1889, 98, 120
Grimod de La Reynière, Alexandre, 1, 42
Grossard, Dom, 28
"Guerre des Deux Haricots, La." *See* "War of the Two Beans, The"

Hau, Michel, 52, 112
Haute-Marne, department of, 5, 48, 119, 133, 147, 166
Haute Montagne, 197
Hautvillers, 48, 125, 155, 197, 231; Abbey, 2, 28, 29, 147; Berthet of, 159
Heidsieck firm, 14, 22, 72, 75, 164
Hérault, 52, 89
Herriot, Édouard, 136
Heuser, Mme., 71, 73
Histoire de la vigne et du vin en France des origines au XIXe siècle (Dion), 2–3, 231
Humanité, L', 176
hybrids, vine, 192

IGP. *See* indication géographique protégée
INAO. *See* Institut national des appellations d'origine
Indépendant rémois, L', 109
indication géographique protégée (IGP), 121
insecticides. *See* pesticides and fungicides
Institut national des appellations d'origine (INAO), 192–93

Jacquot firm, 172
Jaurès, Jean, 82, 136, 160, 168, 169, 183
Joan of Arc, 34, 35
Journal d'agriculture pratique, 93
Jura, 62

Kessler, George, 33
Koch firm, 22, 75
Königsberg, 15
Krug, Paul, 131, 133–34, 156, 164, 235
Kunkelmann firm, 22, 72, 75, 104

labels, champagne, 18–19, 27, 30, 33–35
labor/labor unions, 49, 53, 130, 160, 171, 178–82, 183. See also wage labor
labor exchange. See bourse du travail
Laboratorie municipal (Paris), 106
Lamarre, René, 80, 81, 88, 112, 116, 123, 129, 143, 182, 194, 237; accusations about lack of authenticity in champagne industry, 45–46, 86–87; on delimitation, 126–27; encouragement of peasant resistance by, 108–9; idea for locally based peasant syndicates, 111, 114, 117; and *La Révolution champenoise*, 40–41, 82–85, 99, 101–2, 104–5, 121, 123
Languedoc, 49, 52
Laurent, Robert, 47
Lavisse, Ernest, 136
Law of June 15, 1878, 99
Law of 1905, 135, 140–41, 166, 191
Leech, John, 37
Lehning, James, 185
Lemaire firm, 172
Leroy-Beaulieu, Paul, 125
Ligue des patriotes, 173, 187
"Livre noir des assassins de la Champagne," 170–75, 237
Loire, 97, 140
Lombard, Léandre-Moïse, 1
Lorraine, 63, 83, 121, 122, 187
Lorson, Daniel, 193, 194
Loubère, Leo, 18, 49, 122
louée locale (local hiring fair), 63
Louise of Savoy, 40
Louis XV, 14
Ludes, 48, 64, 197
Luxembourg, 79
Lyons, 20

Mailly-Champagne, 197
Maine-et-Loire, 122
Marchand, Roland, 30

Mareuil, 55, 110, 174, 197
marketing of champagne, 15–20, 27, 29–39, 44, 188, 236
Marne, division of vineyards in, 197
Marne, prefect of, 153, 154, 164
Marne Fédération, 171
Marseille, 152
Marx, Karl, 129
Matin, Le, 151
"Medicinal Champagne," 32
Menudier firm, 172
Mercier, Eugène, 37, 236
Mercier firm, 35, 79–80, 96, 164, 170, 171, 174, 188
Mesnil-sur-Oger, 54, 155, 197
méthode rurale, 68
mévente (slump in wine sales), 124, 125, 126, 132, 178
Michelet, Jules, 47, 137, 191
Michel-Lechacheur, Émile, 148, 151, 172–73, 177, 184, 187
Midi, 52, 152, 155, 158, 161, 166, 175; economic crisis and peasant unrest in, 141, 143, 150, 153, 163, 173; fraud in, 119–21; phylloxera in, 92, 98; use of American rootstock in, 107
mildew. See vine diseases
Miller Brewing Company, 44–45
ministry of agriculture, 89, 136, 140–41, 144, 146, 149, 152, 167, 191, 234, 236
ministry of commerce, 89, 96, 120, 234
ministry of foreign affairs, 96, 120
ministry of the interior, 159, 163, 166
Moët, Auban, 56
Moët, Claude, 14
Moët & Chandon firm, 33, 48, 56–57, 70, 73, 84, 94, 102, 107, 115, 116, 140, 170–71
montagne de Reims, 54, 55, 197
Montebello, Lannes de, 154, 163, 174
Montpellier, 89
Moreau, Émile, 142, 143
Moreau, Jules, 142, 150, 154, 178–79, 183
Moussy, 174, 197
Mumm, G. H. (firm), 22, 23, 36, 71, 73–75, 77, 96, 104
Mumm, Hermann, 73
Mumm, Peter Arnold, 73
Mun, Bertrand de, 168, 177

Narbonne, 143
nationalism, 34, 84–85, 130, 181, 184, 190, 193–94, 228n. 138
Naudin, Charles, 90

Navlet, 37
négociants, definition of, 5
Nollevalle, Jean, 168
Nora, Pierre, 145, 189

Oger, 49, 54, 111, 197
oidium. *See* vine diseases

Pams, Jules, 167
Panama Scandal, 67
Paris, 7, 32, 42, 51, 67, 90, 106, 125, 131, 137, 163, 167, 176; consumer cooperatives in, 84, 116; Convention of, 96; debate on delimitations in, 155, 163–64, 168–69, 177; Exposition universelle in, 28, 35, 37; ineffectiveness of government of, 154; restaurants of, 31, 44, 140; wine markets in, 12–13, 49, 52
Parti ouvrier français, 84, 116
Pasteur, Louis, 135
Paysans, Les (Balzac), 43
Péchadre, A., 131–35, 141, 144, 150, 154, 156, 163
Peguy, Charles, 136
Père Goriot (Balzac), 21
Perignon, Dom Pierre, 2, 3, 28–29, 32, 45, 48, 125, 147, 148, 231
Perrier firm, 158–59, 162, 172
Perrin, Alphonse, 126–27, 129, 130, 143, 151, 152
pesticides and fungicides, 50, 57, 113
Petit Journal, 29
Petit manuel and calendrier phylloxériques à l'usage des vignerons de Champagne, 93
Philbert, Victor, 143, 152, 153
phylloxera, 55, 56, 60, 73, 85, 86–117, 119–21, 123, 125–26, 132, 134, 139, 152, 156, 187, 206n. 39; in Aube, 167; in Midi, 81
Pierry, 174, 176, 197
Piper family, 72
Planchon, Jules-Émile, 89–90, 91
Plankaert firm, 40, 84
planting: *en foule*, 55, 61–62, 113, 114; *en ligne*, 113, 114
Pochet, Marcel, 135
Poincaré, Raymond, 39, 186
Poittevin, Gaston, 143, 152, 155, 161
Polignac, the marquis de, 125, 164
Pommery firm, 174, 236
Pompadour, Madame de, 14, 37
Prestwich, Patricia, 162
privilégiés (purveyors by appointment), 18
Prussia, 14, 43
Pyrenees, 190, 234

Radical party, 111, 129, 131, 149, 176, 181; on delimitations, 169; newspaper of, 109; themes of, 67; vignerons losing faith in, 161
Reddy, William, 21
Reims, 16, 19, 49, 56, 73, 113, 123, 133, 135, 146, 154, 172, 180, 197; antipathy to newcomers in, 20; and *La Champagne commerciale*, 165; legal proceedings in, 153; Moët and Chandon in, 70; phantom firms in, 97; population growth in, 53; and revolt of 1911, 163–64, 171, 176; success of wine merchants in, 13, 15; and Veuve Pommery, 72; Werlé in, 22; wine cellar workers in, 128, 142
Renan, Ernest, 3
Rennes, 67
resistance, peasant: before the Revolution, 47; in 1892, 104–10; in 1894, 111–12; in 1908, 150; in late 1910, 154; in 1911 (*see* revolt of 1911); in 1912–1914 (strike actions), 178–85
Restoration, 14
Réveil de la Marne, 151, 154
revolt of 1911, 7, 66, 143, 156, 158–85
Révolution champenoise, La, 40, 46, 82–85, 99, 101, 104, 237; conversion of, into weekly newspaper, 102; on fraud, 121, 123; list of protest candidates in, 105; phylloxera discussed in, 110
Revue de Paris, 155
Revue des deux mondes, 36
Revue de viticulture, 122
Rhine region, 173
Riley, C. V., 91
Rilly-la-Montagne, 197
Roederer firm, 23, 40, 73, 84
Rolland, Romain, 136–37
Rome, 1886 conference in, 96
Rondeau firm, 174
roquefort cheese, 19, 44, 192
Rouen, 20
Ruinart family, 14
Russia, 14–15, 18, 37, 78, 99, 235

Sahlins, Peter, 190
Sahut, Felix, 89
Saillet, Jacqueline, 176
Sainte-Ménéhould, 51
Saint-Julien, 27
Saint-Pierre, 28
Saumur, 97, 122
Second Empire, 23, 39, 49, 166

second fermentation, 49, 70, 79
Segal, Aaron, 39
Seine-et-Marne, 92, 119
Sièyes, Abbé, 41
Sillery, 27, 55, 80, 97, 197
Six semaines en pays phylloxérés, 94
Société d'agriculture, de commerce, des sciences et des arts de la Marne, 92
socialists, 84, 116, 160, 169, 183
Solidarité (Bourgeois), 129
Sorbonne, 136
Spain, 18, 40, 131
Spang, Rebecca, 140
Sternhell, Zeev, 181
strikes. *See* resistance, peasant
"Swiss method," 94, 95, 107
Switzerland, 14, 22, 50, 94, 131, 193
Syndicat commercial viticole champenois, 142
Syndicat contre le phylloxéra, 95
Syndicat de débitants de boissons, 153
Syndicat de défense des intérêts des négociants en vins de Champagne, 165, 167–68, 177–78, 180
Syndicat des grandes marques de Champagne, 2, 235
Syndicat des négociants en vin de Champagne, 78
Syndicat des ouvriers champenois, 143
Syndicat du commerce des vins de Champagne, 12, 45, 85, 163; advertising by, 37; attempts at delimitation by, 25, 27, 79–80, 122, 123, 125, 126, 131–35, 141–43, 149–52, 156–57, 163–65, 167–68, 177, 180–81; and the AVC, 115; formation of, 26, 77–78, 81; and grape prices, 127, 130; immigrants in membership of, 24; in lawsuit against Miller Brewing Co., 44–45, 235; membership lists of, 71; and the phylloxera crisis, 92, 95–97, 99–102, 106–7, 109; records of, 58, 82, 234; and strike action, 180
Syndicat viticole d'Épernay, 74

Tableau de la géographie de France (Vidal de la Blache), 43, 137–38
tâcherons, 56, 57
Temps, Le, 141
Tennis Court Oath, 34
terroir, definition of, 2, 41–43, 119, 128–29
textile industry, 20, 49, 53, 75, 77
Third Republic, 39, 65, 148; collective action within, 117; and delimitations, 161, 168–69, 182; and Dreyfus Affair, 84; electoral support for, 88; and fraud, 119, 120; and geography, 138; and nationalism, 190–91; and phylloxera, 90; population decline during, 53; and Radical party, 130; and reestablishment of ancien régime privileges, 149; and rural France, 66, 190, 233
Tocqueville, Alexis de, 149
Tomes, Robert, 16, 22
Tour de la France par deux enfants, Le (Fouillée), 139
Tovey, Charles, 17, 20, 71
travail fou (or *travail à quatre bras*), 93, 177
Tréloup, 73, 93
Troyes, 53, 54, 135, 147, 169

United States, 33, 77, 78, 121, 193, 235; consumption of champagne in, 210n. 131; phylloxera in, 89. *See also* American rootstock

Vallé, Paul, 84, 116
Van Cassel firm, 174
Vauciennes, 109
veillées, 65
Venoge firm, 18
Venteuil, 60, 109, 127, 158, 159, 163
Vertus, 55, 111, 112, 125, 197
Verzenay, 27, 48, 49, 73, 128, 180, 197
Verzy, 49, 125, 197
Veuve Clicquot (wine), 96, 140
Veuve Clicquot–Ponsardin (firm), 20, 22, 23, 25, 72, 73, 113, 125, 237
Veuve Damas, 35
Veuve Pommery, 72, 73
Victoria Wine Company, 36
Vidal de la Blache, Paul, 43–44, 136–39, 144, 155, 161, 191
Viger, Albert, 111
Vigilance Committee, 60, 91–92, 93, 94
Vigne, La, 59
vigneron, definition of, 5
vigneron champenois, Le, 78, 81, 104–5, 109, 118, 237; on fraud, 99; on phylloxera, 93–94; on the Syndicat du commerce, 101
Viguié, P., 95, 100, 102, 103, 105, 108
Vimont, Georges, 105, 107, 108, 111; campaign of, against chemical treatment for phylloxera, 95, 101–2, 110; advocacy for locally based peasant syndicats, 60, 91–93, 114, 117

Vinay, 176, 197
Vincelles, 101, 108
Vincennes, 152
Vincent, Saint, 65; feast of, 163
Vin de Champagne, Le, 28
vine diseases, 5, 50, 60, 62, 113, 150, 153. *See also* phylloxera
Viztelly, Henry, 58, 74, 97
Vosges, 130

wage labor, 56, 57, 142, 167, 178–79
Walbaum, Florens, 77, 107
Walbaum firm, 22, 72, 75
Warner, Charles, 119, 124
"War of the Two Beans, The," 144–45, 146, 155

Weber, Eugen, 124, 128–29
Weber, Max, 19
Werlé, Alfred, 108, 109, 115
Werlé, Mathieu-Édouard, 22
Werlé family, 72
What Is the Third Estate? (Sièyes), 41
Wine Trade Review, 33
women in the champagne industry, 34–35, 37, 58, 66
World War I, 5, 11, 23, 33, 38, 139, 178; and anti-German feeling, 181; Bolo-Pacha's role in, 152; effects of, on vineyards, 186–87; and patriotism, 34
World War II, 136